Assessing the Hazard of Metals and Inorganic Metal Substances in Aquatic and Terrestrial Systems

Other Titles from the Society of Environmental Toxicology and Chemistry (SETAC):

Use of Sediment Quality Guidelines and Related Tools for the Assessment of Contaminated Sediments
Wenning, Batley, Ingersoll, Moore, editors
2005

Ecological Assessment of Aquatic Resources: Linking Science to Decision-Making
Barbour, Norton, Preston, Thornton, editors
2004

Amphibian Decline: An Integrated Analysis of Multiple Stressor Effects
Linder, Krest, Sparling, editors
2003

Metals in Aquatic Systems:
A Review of Exposure, Bioaccumulation, and Toxicity Models
Paquin, Farley, Santore, Kavvadas, Mooney, Winfield, Wu, Di Toro
2003

Silver: Environmental Transport, Fate, Effects, and Models:
Papers from Environmental Toxicology and Chemistry, 1983 to 2002
Gorusch, Kramer, La Point, editors
2003

Contaminated Soils: From Soil–Chemical Interactions to Ecosystem Management
Lanno, editor
2003

Environmental Impacts of Pulp and Paper Waste Streams
Stuthridge, van den Heuvel, Marvin, Slade, Gifford, editors
2003

Porewater Toxicity Testing: Biological, Chemical, and Ecological Considerations
Carr and Nipper, editors
2003

Reevaluation of the State of the Science for Water-Quality Criteria Development
Reiley, Stubblefield, Adams, Di Toro, Erickson, Hodson, Keating Jr, editors
2003

For information about SETAC publications, including SETAC's international journals, *Environmental Toxicology and Chemistry* and *Integrated Environmental Assessment and Management*, contact the SETAC Administrative Office nearest you:

SETAC Office
1010 North 12th Avenue
Pensacola, FL 32501-3367 USA
T 850 469 1500   F 850 469 9778
E setac@setac.org

SETAC Office
Avenue de la Toison d'Or 67
B-1060 Brussels, Belgium
T 32 2 772 72 81   F 32 2 770 53 86
E setac@setaceu.org

www.setac.org

Environmental Quality Through Science®

Assessing the Hazard of Metals and Inorganic Metal Substances in Aquatic and Terrestrial Systems

Edited by
William J. Adams
Peter M. Chapman

Proceedings from the Workshop on
Hazard Identification Approach for Metals and
Inorganic Metal Substances

3-8 May 2003
Pensacola Beach, Florida USA

Coordinating Editor of SETAC Books
Joseph W. Gorsuch
Gorsuch Environmental Management Services, Inc.
Webster, New York, USA

CRC Press
Taylor & Francis Group
Boca Raton London New York

CRC Press is an imprint of the
Taylor & Francis Group, an **informa** business

CRC Press
Taylor & Francis Group
6000 Broken Sound Parkway NW, Suite 300
Boca Raton, FL 33487-2742

First issued in paperback 2019

ISBN-13: 978-1-880611-89-0 (SETAC Press)
ISBN-13: 978-1-4200-4440-9 (hbk)
ISBN-13: 978-0-367-38955-0 (pbk)

Library of Congress Cataloging-in-Publication Data

Adams, William J., 1946-
   Assessing the hazard of metals and inorganic metal substances in aquatic and terrestrial systems / William J. Adams and Peter M. Chapman.
      p. cm.
   Includes bibliographical references and index.
   ISBN-13: 978-1-4200-4440-9 (alk. paper)
   1. Metals--Environmental aspects. 2. Environmental risk assessment. 3. Metals--Toxicology. I. Chapman, Peter M. II. Title.

TD196.M4A33 2006
577.27--dc22
2006022030

Visit the Taylor & Francis Web site at
http://www.taylorandfrancis.com

and the CRC Press Web site at
http://www.crcpress.com

SETAC Publications

Books published by the Society of Environmental Toxicology and Chemistry (SETAC) provide in-depth reviews and critical appraisals on scientific subjects relevant to understanding the impacts of chemicals and technology on the environment. The books explore topics reviewed and recommended by the Publications Advisory Council and approved by the SETAC North America Board of Directors, SETAC Europe Council, or SETAC World Council for their importance, timeliness, and contribution to multidisciplinary approaches to solving environmental problems. The diversity and breadth of subjects covered in the publications reflect the wide range of disciplines encompassed by environmental toxicology, environmental chemistry, hazard and risk assessment, and life-cycle assessment. SETAC books attempt to present the reader with authoritative coverage of the literature, as well as paradigms, methodologies, and controversies; research needs; and new developments specific to the featured topics. The books are generally peer reviewed for SETAC by acknowledged experts.

SETAC publications, which include Technical Issue Papers (TIPs), workshop summaries, a newsletter (*SETAC Globe*), and journals (*Environmental Toxicology and Chemistry* and *Integrated Environmental Assessment and Management*), are useful to environmental scientists in research, research management, chemical manufacturing and regulation, risk assessment, life-cycle assessment, and education, as well as to students considering or preparing for careers in these areas. The publications provide information for keeping abreast of recent developments in familiar subject areas and for rapid introduction to principles and approaches in new subject areas.

Table of Contents

Chapter 6 Hazard Assessment of Inorganic Metals and Metal Substances in Terrestrial Systems 113

Erik Smolders, Steve McGrath, Anne Fairbrother, Beverley A. Hale, Enzo Lombi, Michael McLaughlin, Michiel Rutgers, and Leana Van der Vliet

Appendix A: A Unit World Model for Hazard Assessment of Organics and Metals 135

Acknowledgments

This book presents the proceedings of a Pellston Workshop convened by the Society of Environmental Toxicology and Chemistry (SETAC) in Pensacola, Florida, in May 2003. The 47 scientists, managers, and policymakers involved in this workshop represented seven countries. We thank all participants for their contributions, both in the workshop and in subsequent discussions resulting in this book.

The workshop and this book were made possible by the generous support of the following organizations (in alphabetical order):

- Center for the Study of Metals in the Environment (CSME)
- Environment Canada
- Eurometaux
- International Copper Association
- International Lead Zinc Research Organization
- Kennecott Utah Copper Corporation
- Kodak
- Natural Resources Canada
- Nickel Producers Environmental Research Association (NiPERA)
- Rio Tinto
- U.S. Environmental Protection Agency (Office of Research and Development)

The workshop would also not have been possible without the very capable management and excellent guidance provided by Greg Schiefer, Linda Longsworth, and Mimi Meredith, and the support of SETAC Executive Director Rodney Parrish. In particular, the efforts of Mimi Meredith in the production of this book are gratefully acknowledged.

William J. Adams
Peter M. Chapman

Editors

William J. Adams, Ph.D. is a Principal Environmental Scientist and General Manager for Rio Tinto, Salt Lake City, Utah. He was previously the Director of Environmental Science for 6 years at Kennecott Utah Copper, Vice President of ABC Laboratories for 5 years, and Science Fellow at Monsanto Company for 14 years. His research interests include developing ecotoxicology risk assessment methods for metals, site-specific methodologies for water quality criteria for metals, and development of an approach for hazard assessment of metals. Dr. Adams has published several papers on methods for assessing sediments and was instrumental in developing the science supporting equilibrium partitioning theory (EqP) for nonpolar organic substances. He has also published in the area of water quality assessments. He was a member of the U.S. Environmental Protection Agency (EPA) Science Advisory Board (SAB) for 10 years and has served on several other national committees.

Peter M. Chapman is a Principal and Senior Environmental Scientist with Golder Associates in North Vancouver, British Columbia, Canada. He has been an active researcher for almost 30 years in the fields of aquatic ecology, ecotoxicology, and environmental risk assessment, with a particular focus on metals and metalloids. He has published more than 140 articles in international, peer-reviewed scientific journals, and in book chapters. He is Senior Editor of the international, peer-reviewed journal *Human and Ecological Risk Assessment*, a member of the editorial board of two other international peer-reviewed journals, and edits a highly popular series of scientific "Learned Discourses" in the *SETAC Globe*. In 1996 he received an award from the EPA for resolving environmental issues in Port Valdez, Alaska. In 2001, the Society of Environmental Toxicology and Chemistry awarded him their highest award for lifetime achievement and outstanding contributions to the environmental sciences: The Founders Award.

Workshop Participants*

Steering Committee Member (SCM)
Workshop Chair (WC)

WORKGROUP 1: PERSISTENCE

William J. Adams (WC) (SCM)
Rio Tinto
Murray, Utah

William Davison
Lancaster University
Lancaster, United Kingdom

Dominic M. Di Toro
Hydroqual
Englewood, New Jersey

Miriam L. Diamond
University of Toronto
Toronto, Ontario, Canada

Patrick J. Doyle (SCM)
Environment Canada
Hull, Ontario, Canada

Samuel Luoma
U.S. Geological Survey
Menlo Park, California

Donald Mackay
Trent University
Peterborough, Ontario, Canada

Jerome Nriagu
University of Michigan
Ann Arbor, Michigan

Johanna Peltola-Thies
Federal Environmental Agency
Berlin, Germany

Adam Peters (Rapporteur)
Environment Agency
Wallingford, United Kingdom

Carol Ptacek
University of Waterloo
Waterloo, Ontario, Canada

James M. Skeaff
Natural Resources Canada
Ottawa, Ontario, Canada

Edward Tipping
Centre for Ecology and Hydrology
Cumbria, United Kingdom

Hugo Waeterschoot
Eurometaux
Brussels, Belgium

John Westall (Chair) (SCM)
Oregon State University
Corvallis, Oregon

William Wood (SCM)
U.S. Environmental Protection Agency
Washington, D.C.

* Affiliations were current at the time of the workshop.

WORKGROUP 2: BIOACCUMULATION

Ronny Blust
University of Antwerp
Antwerp, Belgium

Uwe Borgmann
Environment Canada
Burlington, Ontario, Canada

Kevin Brix
EcoTox
Newport, Oregon

Nicolas Bury
King's College
London, United Kingdom

Yves Couillard
Environment Canada
Hull, Quebec, Canada

Robert L. Dwyer
International Copper Association
New York, New York

Samuel N. Luoma
U.S. Geological Survey
Menlo Park, California

James C. McGeer (Rapporteur)
Natural Resources Canada
Ottawa, Ontario, Canada

Steve Robertson (SCM)
Environment Agency
Wallingford, United Kingdom

Keith G. Sappington
U.S. Environmental Protection Agency
Washington, D.C.

Christian Schlekat (Chair)
U.S. Borax
Valencia, California

Ilse Schoeters (SCM)
European Copper Institute
Brussels, Belgium

Dick T.H.M. Sijm
National Institute for Public Health and
the Environment (RIVM)
Bilthoven, The Netherlands

WORKGROUP 3: TOXICITY

Herbert E. Allen
University of Delaware
Newark, Delaware

Peter G.C. Campbell (SCM)
Université du Québec
Ste. Foy, Quebec, Canada

Richard D. Cardwell
Parametrix
Corvallis, Oregon

**Peter M. Chapman (Rapporteur)
(SCM)**
EVS Consultants
North Vancouver, British Columbia,
Canada

Amy Crook
Center for Science in Public
Participation/Environmental Mining
Council
Victoria, British Columbia, Canada

Karel De Schamphelaere
University of Ghent
Ghent, Belgium

Katrien Delbeke
European Copper Institute
Brussels, Belgium

Andrew S. Green (Chair)
International Lead Zinc Research
  Organization
Research Triangle Park, North Carolina

David R. Mount
U.S. Environmental Protection Agency
Duluth, Minnesota

William A. Stubblefield
Parametrix
Corvallis, Oregon

WORKGROUP 4: TERRESTRIAL

Anne Fairbrother
U.S. Environmental Protection Agency
Corvallis, Oregon

Beverly A. Hale
University of Guelph
Guelph, Ontario, Canada

Enzo Lombi
Commonwealth Scientific and
  Industrial Research Organization
  (CSIRO) Land and Water
Adelaide, Australia

Steve McGrath
IACR-Rothamsted
Herts, United Kingdom

Michael McLaughlin
Commonwealth Scientific and
  Industrial Research Organization
  (CSIRO) Land and Water
Adelaide, Australia

Michiel Rutgers
National Institute for Public Health and
  the Environment (RIVM)
Bilthoven, The Netherlands

Erik Smolders (Chair) (SCM)
K.U. Leuven
Heverlee, Leuven, Belgium

Leana Van der Vliet
Environment Canada
Ottawa, Ontario, Canada

Other SETAC Books

Perchlorate Ecotoxicology
Kendall and Smith, editors
2006

Natural Attenuation of Trace Element Availability in Soils
Hamon, McLaughlin, Stevens, editors
2006

Mercury Cycling in a Wetland-Dominated Ecosystem: A Multidisciplinary Study
O'Driscoll, Rencz, Lean
2005

Atrazine in North American Surface Waters:
A Probabilistic Aquatic Ecological Risk Assessment
Giddings, editor
2005

Effects of Pesticides in the Field
Liess, Brown, Dohmen, Duquesne, Hart, Heimbach, Kreuger, Lagadic, Maund,
Reinert, Streloke, Tarazona
2005

Human Pharmaceuticals: Assessing the Impacts on Aquatic Ecosystems
Williams, editor
2005

Toxicity of Dietborne Metals to Aquatic Organisms
Meyer, Adams, Brix, Luoma, Stubblefield, Wood, editors
2005

Toxicity Reduction and Toxicity Identification Evaluations for Effluents, Ambient
Waters, and Other Aqueous Media
Norberg-King, Ausley, Burton, Goodfellow, Miller, Waller, editors
2005

Use of Sediment Quality Guidelines and Related Tools for the Assessment of
Contaminated Sediments
Wenning, Batley, Ingersoll, Moore, editors
2005

Life-Cycle Assessment of Metals
Dubreuil, editor
2005

Working Environment in Life-Cycle Assessment
Poulsen and Jensen, editors
2005

Life-Cycle Management
Hunkeler, Saur, Rebitzer, Finkbeiner, Schmidt, Jensen, Stranddorf, Christiansen
2004

Scenarios in Life-Cycle Assessment
Rebitzer and Ekvall, editors

Ecological Assessment of Aquatic Resources: Linking Science to Decision-Making
Barbour, Norton, Preston, Thornton, editors
2004

Life-Cycle Assessment and SETAC: 1991–1999
15 LCA publications on CD-ROM
2003

Amphibian Decline: An Integrated Analysis of Multiple Stressor Effects
Greg Linder, Sherry K. Krest, Donald W. Sparling
2003

Metals in Aquatic Systems:
A Review of Exposure, Bioaccumulation, and Toxicity Models
Paquin, Farley, Santore, Kavvadas, Mooney, Winfield, Wu, Di Toro
2003

Silver: Environmental Transport, Fate, Effects, and Models:
Papers from Environmental Toxicology and Chemistry, 1983 to 2002
Gorusch, Kramer, La Point
2003

Code of Life-Cycle Inventory Practice
de Beaufort-Langeveld, Bretz, van Hoof, Hischier, Jean, Tanner, Huijbregts, editors
2003

Contaminated Soils: From Soil–Chemical Interactions to Ecosystem Management
Lanno, editor
2003

Environmental Impacts of Pulp and Paper Waste Streams
Stuthridge, van den Heuvel, Marvin, Slade, Gifford, editors
2003

Life-Cycle Assessment in Building and Construction
Kotaji, Edwards, Shuurmans, editors
2003

Porewater Toxicity Testing: Biological, Chemical, and Ecological Considerations
Carr and Nipper, editors
2003

Reevaluation of the State of the Science for Water-Quality Criteria Development
Reiley, Stubblefield, Adams, Di Toro, Erickson, Hodson, Keating Jr, editors
2003

5th LCA Case Studies Symposium
1997

Atmospheric Deposition of Contaminants to the Great Lakes and Coastal Waters
Baker, editor
1997

Biodegradation Kinetics: Generation and Use of Data
for Regulatory Decision-Making
Hales, Feijtel, King, Fox, Verstraete, editors
1997

Biotransformation in Environmental Risk Assessment
Sijm, de Bruijn, de Boogt, de Wolf, editors
1997

Chemical Ranking and Scoring: Guidelines for Relative Assessments of Chemicals
Swanson and Socha, editors
1997

Chemically Induced Alterations in Functional Development
and Reproduction of Fishes
Rolland, Gilbertson, Peterson, editors
1997

Ecological Risk Assessment of Contaminated Sediments
Ingersoll, Dillon, Biddinger, editors
1997

Life-Cycle Impact Assessment: The State-of-the-Art, 2nd ed.
Barnthouse, Fava, Humphreys, Hunt, Laibson, Moesoen, Owens, Todd, Vigon,
Weitz, Young, editors
1997

Public Policy Application of Life-Cycle Assessment
Allen and Consoli, editors
1997

Quantitative Structure-Activity Relationships (QSAR) in Environmental
Sciences VII
Chen and Sch,rmann, editors
1997,

Reassessment of Metals Criteria for Aquatic Life Protection:
Priorities for Research and Implementation
Bergman and Dorward-King, editors
1997

Simplifying LCA: Just a Cut?
Christiansen, editor
1997

SETAC

A Professional Society for Environmental Scientists and Engineers and Related Disciplines Concerned with Environmental Quality

The Society of Environmental Toxicology and Chemistry (SETAC), with offices currently in North America and Europe, is a nonprofit, professional society established to provide a forum for individuals and institutions engaged in the study of environmental problems, management and regulation of natural resources, education, research and development, and manufacturing and distribution.

Specific goals of the society are:

- Promote research, education, and training in the environmental sciences.
- Promote the systematic application of all relevant scientific disciplines to the evaluation of chemical hazards.
- Participate in the scientific interpretation of issues concerned with hazard assessment and risk analysis.
- Support the development of ecologically acceptable practices and principles.
- Provide a forum (meetings and publications) for communication among professionals in government, business, academia, and other segments of society involved in the use, protection, and management of our environment.

These goals are pursued through the conduct of numerous activities, which include:

- Hold annual meetings with study and workshop sessions, platform and poster papers, and achievement and merit awards.
- Sponsor a monthly scientific journal, a newsletter, and special technical publications.
- Provide funds for education and training through the SETAC Scholarship/Fellowship Program.
- Organize and sponsor chapters to provide a forum for the presentation of scientific data and for the interchange and study of information about local concerns.
- Provide advice and counsel to technical and nontechnical persons through a number of standing and ad hoc committees.

SETAC membership currently is composed of more than 5000 individuals from government, academia, business, and public-interest groups with technical backgrounds in chemistry, toxicology, biology, ecology, atmospheric sciences, health sciences, earth sciences, and engineering.

If you have training in these or related disciplines and are engaged in the study, use, or management of environmental resources, SETAC can fulfill your professional affiliation needs.

All members receive a newsletter highlighting environmental topics and SETAC activities, and reduced fees for the Annual Meeting and SETAC special publications.

All members except Students and Senior Active Members receive monthly issues of *Environmental Toxicology and Chemistry (ET&C)* and *Integrated Environmental Assessment and Management (IEAM),* peer-reviewed journals of the Society. Student and Senior Active Members may subscribe to the journal. Members may hold office and, with the Emeritus Members, constitute the voting membership.

If you desire further information, contact the appropriate SETAC Office.

1010 North 12th Avenue
Pensacola, Florida 32501-3367 USA
T 850 469 1500   F 850 469 9778 86
E setac@setac.org

Avenue de la Toison d'Or 67
B-1060 Brussels, Belgium
T 32 2 772 72 81   F 32 2 770 53

E setac@setaceu.org

www.setac.org

Environmental Quality Through Science®

1 A Pellston Workshop on Metals Hazard Assessment

William J. Adams and Peter M. Chapman

1.1 INTRODUCTION TO THE WORKSHOP

This book is the result of discussions that took place at the Pellston Workshop on Assessing the Hazard of Metals and Inorganic Metal Substances in Aquatic and Terrestrial Systems. The workshop, sponsored by the Society of Environmental Toxicology and Chemistry (SETAC), was held 3–8 May, 2003, in Pensacola, FL. The workshop built upon the findings of a previous SETAC workshop, which provided an in-depth discussion of the potential to assess bioavailability of metals to fish and invertebrates (Bergman and Dorward-King 1996) and which led to the development of the Biotic Ligand Model (BLM) (Di Toro et al. 2001, 2005).

The purpose of the workshop was to allow for a focused discussion regarding the fate and effects of metals in the environment (the focus was on inorganic substances; however, where appropriate, organometallic substances were also considered) and incorporating important advances in the state of knowledge that had occurred in the intervening 7 years. Specifically, this workshop allowed for a forum for further discussions among scientists, environmental regulators, and environmental managers, on the utility of persistence, bioaccumulation, and toxicity (PBT) for hazard identification and classification procedures for metals and inorganic metal substances.

The workshop brought together a multidisciplinary and international group of 47 scientists, managers, and policymakers from Australia, Belgium, Canada, Germany, The Netherlands, the United Kingdom, and the United States for 6 days of discussions on various means to assess the environmental hazard posed by metals and inorganic metal substances. Participants included representatives from regulatory and nonregulatory government agencies, academia, industry, environmental groups, and consulting firms involved in assessment, management, and basic research on metals and metal substances.

During the first day of the workshop, presentations were given on the application of PBT criteria in the different regulatory arenas in Canada, Europe, and the United States. Additional presentations highlighted the state of the science regarding the interpretation of PBT for metals. These presentations provided the basis for subsequent plenary and workgroup discussions. Participants were assigned to 4 different workgroups as follows:

1. Persistence — reviewing the scientific underpinnings of the use of persistence in hazard evaluation and of persistence measures as applied to metals, including the potential to use bioavailability measures in aquatic systems.
2. Bioaccumulation — reviewing the soundness of current uses of bioaccumulation in hazard evaluation of metals in aquatic species and aquatic-linked food chains.
3. Toxicity — reviewing toxicity procedures used to assess the hazard of metals as used within PBT approaches.
4. Terrestrial systems — evaluating current uses of PBT measures for metals in terrestrial ecosystems, with a view to improving the approach or identifying an alternative methodology.

In each of these discussions, participants were urged to seek consensus, where possible, on specific technical issues of concern for assessing the hazard of metals and metal substances, and to identify recommendations for future research that could lead to improvements in the existing methods available. Chapter 3 through Chapter 6 in this book provide a synopsis of the discussions and conclusions from each of the workgroups; an overall executive summary is provided in Chapter 2.

This book provides the basis for substantive improvements to the current model for the hazard assessment of metals and metal substances. It is our hope that this book will not only advance the science, but will also serve as the basis for further discussions and advances in the foreseeable future.

1.2 HAZARD IDENTIFICATION, CLASSIFICATION, AND ASSESSMENT

Hazard identification and classification procedures currently used in many countries are based on PBT measurements. Procedures for aquatic hazard identification or classification of organic and inorganic substances have been harmonized by the Organisation for Economic Cooperation and Development (OECD 2001) for the purpose of classifying market-place substances in terms of their potential hazard. PBT criteria are further used within the regulatory context to rank and identify substances of concern. In the United States, PBT criteria have been used to identify substances of concern for waste minimization, emissions reporting, and for the identification of substances for stricter regulations (air, water, and solid waste). In Canada, a PBT-type approach is used for categorizing substances on the Domestic Substances List (DSL) to determine if a screening assessment is required. Depending upon the assessment findings, actions to reduce exposure may be taken. In the European Union (EU), in the framework of the New Chemicals Policy, discussions are ongoing on whether to use PBT criteria to identify substances of very high concern, which will have to be given use-specific permission before they can be employed in particular uses. In addition, the EU New Chemicals Policy (REACH: Registration, Evaluation, Authorization, and Restriction of Chemicals) will necessitate authorization for use of substances classified as PBT and vPvB (very persistent and very bioaccumulative).

Materials used in manufacturing and commerce may be hazardous to the environment. *Hazard* is defined as a measure of the inherent (intrinsic) capacity of a substance to cause an adverse response in a living organism (OECD 1995). Organisms will be placed at possible risk if the substance enters the environment, with the degree (probability) of risk related to the hazardous nature of the substance and the amount of exposure that occurs. Therefore, substances that are very hazardous have a greater likelihood of causing environmental injury in the case of spills or other accidents than those that are less hazardous. Hazard assessment is differentiated from risk assessment in that it does not quantitatively evaluate exposure and deals with inherent properties, not probabilities. Measures of persistence, such as biodegradation and hydrolysis, may be viewed as surrogates of biota exposure to different substances. There have been several primary uses of hazard information:

- environmental hazard classification of substances;
- ranking and/or selection of priority substances;
- Selection of contaminated sites for further evaluation;
- derivation of water, soil, and sediment quality guidelines or criteria for individual substances; and
- ecological risk assessments, both site-specific (i.e., local) and generic (i.e., regional), in conjunction with appropriate exposure data.

A more detailed discussion on hazard assessment of metals is presented in Adams et al. (2000) and Fairbrother et al. (2002).

The scientific community and many regulators recognize that there are significant challenges associated with the application of traditional PBT hazard evaluation tools for inorganic metals and metal substances (collectively termed metals) and that additional tools and techniques may be needed for the proper hazard identification and risk assessment of metals. Further, it is understood that hazard (and risk) assessment must be performed in such a way as to ensure that all substances are evaluated equally and fairly while ensuring that both the environment and human health are protected.

Key issues associated with the application of PBT concepts to metals are as follows (full details are provided in the respective chapters):

Persistence (Chapter 3): Traditional degradation mechanisms used for organic substances to evaluate persistence (or the converse, biodegradation) of metals have been criticized as inappropriate (Canada/European Union 1996). A key question remains as to whether alternative mechanisms and measurements are needed for metals and, if so, which of these are acceptable and under what conditions do they apply? Although it is recognized that metals are conserved, the form and availability of the metal can change and are different for each metal element.

Bioaccumulation (Chapter 4): Unlike organic substances, bioaccumulation potential of metals cannot be estimated using log octanol–water partition coefficients (Log K_{ow}). Bioconcentration and bioaccumulation factors (BCFs and BAFs) are inversely related to exposure concentration and are not reliable predictors of chronic toxicity or food chain accumulation for most aquatic organisms and most metals (Chapman and Wang 2000). The inverse relationship between exposure concentration and BCF results in organisms from the cleanest environments (i.e., background) having the largest BCF or BAF values. This result is counterintuitive to the use of BCF and log K_{ow} as originally derived for organic substances (McGeer et al. 2003). Many organisms appear to regulate metal accumulation to some extent, especially for essential metals.

Toxicity (Chapter 5): Metals are generally not readily soluble. Toxicity test results based on soluble salts may overestimate the bioavailability and the potential for toxicity for many substances, especially for the massive metals and insoluble sulfide and metal oxide forms.

1.3 WORKSHOP PURPOSE AND GOALS

The purpose of this workshop was to identify limitations in the use of PBT for hazard assessment of metals and propose improvements or alternatives. A series of questions were posed for each working group (WG) as a means to initiate discussion. However, the WGs were not required to answer each question; rather, they were presented with the following challenge: to review the science underpinning the use and measurement of PBT for hazard identification of metals in the aquatic environment, propose alternatives or improvements, and identify a hazard assessment approach for terrestrial ecosystems. It was recognized that the development of an integrated approach for hazard assessment would present the best outcome, provided such an approach could be developed. In fact, such an approach, termed the unit world model (UWM) was developed and is presented in detail in Chapter 3.

REFERENCES

Adams WJ, Conard B, Ethier G, Brix KV, Paquin PR, DiToro DM. 2000. The challenges of hazard identification and classification of insoluble metals and metal substances for the aquatic environment. Human Ecol Risk Assess 6:1019–1038.

Bergman HL, Dorward-King EJ. 1996. Reassessment of metals criteria for aquatic life protection. Pensacola, FL: SETAC Press.

Canada/European Union. 1996. Technical Workshop on biodegradation/persistence and bioaccumulation/biomagnification of metals and metal compounds. Brussels, Belgium.

Chapman PM, Wang F. 2000. Issues in ecological risk assessment of inorganic metals and metalloids. Human Ecol Risk Assess 6:965–988.

Di Toro DM, Allen HE, Bergman H, Meyer JS, Paquin PR, Santore CS. 2001. Biotic ligand model of the acute toxicity of metals. 1. Technical basis. Environ Toxicol Chem 20:2383–2396.

Di Toro DM, McGrath JA, Hansen DJ, Berry WJ, Paquin PR, Mathew R, Wu KB, Santore RC. 2005. Predicting sediment metal toxicity using a sediment Biotic Ligand Model: methodology and initial application. Environ Toxicol Chem 24:2410–2427.

Fairbrother A, Glazebrook PW, van Straalen NM, Tarazona JV (eds). 2002. Test methods for hazard determination of metals and sparingly soluble metal compounds in soils. Pensacola, FL: SETAC Press.

McGeer JC, Brix KV, Skeaff JM, DeForest DK, Brigham SI, Adams WJ, Green A. 2003. Inverse relationship between bioconcentration factor and exposure concentration for metals: implications for hazard assessment of metals in the aquatic environment. Environ Toxicol Chem 22:1017–1037.

OECD (Organisation for Economic Cooperation and Development). 1995. Test methods for hazard and risk determination of metals and inorganic metal compounds. Paris, France: OECD.

OECD (Organisation for Economic Cooperation and Development). 2001. harmonized integrated hazard classification system for human health and environmental effects of chemical substances. Available from: http://www.oecd.org/ehs/Class/HCL6.htm.

2 Executive Summary

William J. Adams and Peter M. Chapman

2.1 INTRODUCTION

Current approaches for hazard identification and classification of substances introduced into the environment are largely based upon persistence, bioaccumulation, and toxicity (PBT) measurements. However, there are problems with the application of PBT to metals and metal substances. Persistence and bioaccumulation, as presently formulated, frequently do not adequately consider important metal physicochemical considerations such as speciation, complexation, precipitation, dissolution, transformation, and sedimentation. Further, toxicity, as presently formulated, frequently does not adequately consider bioavailability and too often uses the lowest acceptable toxicity value instead of an integrated approach such as a species sensitivity distribution.

This book reports the findings of a workshop organized around the constructs of PBT for purposes of examining strengths and weakness in each of these criteria and identifying alternatives or improvements that could be recommended for metals and metal substances. Consensus was reached at the workshop that the individual PBT criteria are limited in their ability to assess hazard or to prioritize substances. The criteria are not linked or integrated and they attempt to identify or predict effects (hazard) using bioaccumulation and persistence as modifiers of toxicity, without fully incorporating other important metal fate characteristics.

The primary recommendation from this workshop is that a critical load modeling approach, termed the unit world model (UWM), which integrates appropriate components of PBT into a consolidated modeling approach be used for hazard assessment of metals and metal substances for purposes of ranking or prioritization. The use of a UWM approach is desirable because it is applicable to both metals and organic substances and would allow for comparison of the hazards posed by both classes of substances.

2.2 PERSISTENCE

The UWM approach estimates the rate at which a metal or metal substance can enter a given ecosystem (the unit world) before reaching a concentration (at steady state or after a defined period of time) in one of the compartments of the ecosystem (water, sediment, or soil) that causes effects to biota. Such an approach integrates metal environmental chemistry and fate to estimate critical loads that potentially cause toxic effects. The output of such an approach is an estimate of load of the amount of metal substance required (load/mass, e.g., kg/d) to result in an effect in the model system. This approach accounts for differences in the physicochemical properties between different metals and metal substances and provides a means to identify hazard as a function of the load to the system. A substance with a small critical load would be more hazardous than one with a larger critical load. Hence, this methodology can be used for ranking and prioritization, within the limitations of the modeling approach. Some existing models are capable of performing the necessary calculations to derive critical load estimates; other models are being developed.

2.3 BIOACCUMULATION

The potential for metals bioaccumulation to cause dietary toxicity is included in the UWM, not via inappropriate bioaccumulation and bioconcentration factors (BAFs and BCFs), but rather by means of a comparison of the results of a bioaccumulation submodel to dietary threshold values. Such a submodel ensures that the environmental hazard of metals is not underestimated by ignoring bioaccumulation through the food web, which may cause adverse effects at concentrations below chronic criteria/guideline values. The food web submodel estimates metal concentrations within the tissues of a representative prey organism that result from a given waterborne metal concentration. These tissue concentrations then serve as the exposure concentrations for upper trophic level predators.

Within the UWM framework, if the predicted tissue concentration in the prey organism at the water quality criterion/guideline is less than the dietary threshold for the consumer organism, then dietary toxicity does not represent the limiting pathway with respect to environmental hazard; rather, the overall hazard of the substance will be determined by toxicity thresholds based on direct toxicity to aquatic life. On the other hand, if the predicted tissue concentration in the prey organism at the water quality criterion/guideline exceeds the dietary threshold for the consumer organism, then dietary toxicity is the limiting pathway, and a back calculation to the appropriate safe concentration in water or sediment must be made for use in the UWM framework.

2.4 TOXICITY

Three principles were set forth to ensure that robust and reliable toxicity data are applied in the UWM in relevant environmental compartments. First, test conditions should be normalized (e.g., similar temperatures) and described. Second, the same measurement endpoints should be used (ideally, survival, growth, and fecundity, which reflect population-level effects). Third, toxicity should be reported in terms of comparable metrics (for example, preferably, EC_x values). In addition:

- Data should be screened for quality before use in categorization. Data recognized as having "fatal" shortcomings should be rejected outright. Other data should be categorized as "acceptable" or "interim," depending on their quality. Similar qualifications apply to categorizations based on those data.

- The lowest available toxicity value should not be used to define hazard when an integrative approach is possible. Standardized approaches that normalize data sets based on data quality should be used.
- The water quality from which the test organisms were captured, cultured, and tested should be defined and should be similar to the test medium, with no deficiencies or excesses of essential metals.
- For categorization of metal hazards in sediments, pore water metal concentrations can be used in conjunction with aquatic toxicity values derived from tests of water column and benthic organisms.
- Bioavailability should be used to normalize data sets, reducing uncertainty and increasing comparability when possible.
- Dietary uptake can be a major source of metal body burden for some metals. However, the bioreactivity of inorganic metals within aquatic organisms remains poorly understood. There is presently no clear evidence that water quality guidelines are not protective for both water and dietary exposures to inorganic metals.
- Until the UWM is fully developed, categorization of metals based on toxicity should rely on integration of toxicity and solubility data, based ideally on free metal ion concentrations, or less ideally, on dissolved metal concentrations.

2.5 TERRESTRIAL ENVIRONMENT

Soils are important sinks for metals in the environment. The major routes of metal input to soils are atmospheric deposition, application of sewage sludges, animal manures, inorganic fertilizers, and alluvial deposition. Metals generally have a greater level of adverse effects on biota in aquatic systems than in terrestrial systems over the short term because, in terrestrial systems, metals are bound to soils and, over time, following deposition, their bioavailability decreases markedly.

Hazard ranking of metals in soil depends on the soil type and the toxicological pathways considered, that is, direct toxicity or considerations of secondary poisoning. Hazard ranking is possible using existing soil quality criteria/guidelines from various countries, but significant variation in relative rankings is evident. Also, most of these values are based on direct toxicity pathways, so that ranking using an average value across jurisdictions does not give equal weight to secondary poisoning issues. Further, comparison of hazard ranking using soil quality criteria/guidelines often does not correlate with hazard ranking in a single soil with a single test, so that ranking depends on the critical pathways considered (mammalian, microbial, plant, etc.).

A better ranking system, and one that could be incorporated into the UWM, would involve actual toxicity tests using 3 different trophic levels under set conditions in the laboratory. Such testing should include, at a minimum, 3 specific trophic levels: plants; invertebrates; and microbes. These 3 trophic levels represent primary producers, consumers, and decomposers, which are the key elements of the soil ecosystem.

Three parallel toxicity tests would be performed, the first after a short equili-bration time (7 days). The remaining 2 tests would be performed after a prolonged equilibration time (60 days) with and without a leaching step after 7 days to remove the toxicity of counterions released during dissolution. Testing should involve 2 soils, one that accentuates the bioavailability of cationic metals (pH 5 to 5.5) and the other that maximizes the bioavailability of anions (pH 7.5 to 8). The output generated would be conservative because it is a reasonable worst-case for the 2 forms of ions, allowing for transformations of insoluble compounds. Ideally, hazard assessment would include such toxicity testing in a weight of evidence assessment that also incorporates potential for secondary poisoning of predators.

2.6 CONCLUSION

Development of the UWM was not foreseen as an outcome before the workshop; improvements to the PBT concepts were envisioned. However, the UWM approach was a logical development during the workshop. The UWM comprises an integrated approach to assessing the hazard (and risk) posed by metals and metal substances in the environment. It allows for a continuum of assessments, including evaluations for classification, ranking, and screening, and can be used for both metals and organic substances.

3 Integrated Approach for Hazard Assessment of Metals and Inorganic Metal Substances: The Unit World Model Approach

Adam Peters, William J. Adams,
Miriam L. Diamond, William Davison,
Dominic M. Di Toro, Patrick J. Doyle,
Donald Mackay, Jerome Nriagu,
Carol Ptacek, James M. Skeaff, Edward Tipping,
and Hugo Waeterschoot

3.1 INTRODUCTION

This chapter presents the unit world model (UWM). Subsequent chapters discuss its implementation. The most important feature of this chapter is the synthesis of applicable metal fate and effects concepts into a unifying concept. Efforts to render the UWM a working model rather than simply a unifying concept are underway and will be reported elsewhere.

3.1.1 BACKGROUND

The approach of characterizing the potential hazard of organic chemicals by considering those inherent, chemical-specific properties that relate to their potential persistence, bioaccumulation, and toxicity (P, B, and T, or PBT) in the environment has a long history, with variations having been widely employed throughout the world (EU 1991; Kleka et al. 2000; Lipnick et al. 2000; OECD 2001a; Mackay et al. 2003a). The PBT approach has had wide appeal, at least in part, because it provides a way to address a complex subject in the context of a reasonably well-defined and readily implemented procedure. Given the recognized utility of the PBT approach in the assessment of hazard for some organic chemicals, regulatory agencies have made efforts to apply a similar approach for metals (OECD 2001a; Existing Substances Branch 2003). Although this development has satisfied a clear regulatory need, it has also resulted in the recognition that significant limitations may exist in the application of the PBT approach to metals (Adams et al. 2000), as well as for some types of organic chemicals such as polymers and pigments. The identification of a universally agreed-upon approach to overcome these deficiencies has not been immediately apparent. However, the fact that such deficiencies existed provided the motivation that was needed for a significant effort to be put forth by scientists, regulators, and industry to develop a more refined assessment procedure.

One of the important areas in which the classical PBT approach is deficient is the way it addresses the potential for exposure to chemicals (for example, the exposure concentration). In general, persistence (often expressed as a residence time or as a half-life) serves as a surrogate for exposure information (that is, concentrations in the environment) over long periods of time and over relatively wide areas in a given medium. Persistence, when multiplied by an emission (or input) rate (kg/d), gives the mass (kg) of a chemical in the system. This mass translates into a concentration and, in turn, to a dose from which the potential risk of an adverse effect can be estimated. Persistence can also be used to indicate the potential for a compound to undergo long-range transport to locations far from the point of introduction with subsequent long-term exposure.

Persistence for organic chemicals is generally characterized by the rate at which a chemical is broken down in the environment (for example, by bacterial degradation or photooxidation) into compounds that are typically less hazardous than the original parent compound. For organic compounds that degrade quickly, low persistence is thus related to low potential for exposure. However, for many inorganic chemicals (and some organic chemicals that degrade slowly), other processes that affect the environmental exposure levels are also operative and may be of comparable importance in an evaluation of "potential for exposure." For example, both organic chemicals and metals sorb to particulate material to varying degrees (Di Toro and Paquin 2000; Mackay et al. 2003a), and subsequent settling of this material leads to a decrease in exposure for water-column-based pelagic organisms and an increase in exposure for benthic organisms.

Beyond the manner in which consideration is given to environmental fate via persistence, the fact that the individual P-, B-, and T-related parameters, for both organic chemicals and metals, are often evaluated independently for each environmental compartment also leads to problems of interpretation. This approach misses the linkages that occur in natural systems. As a result, the conclusions that are drawn are often of questionable validity. Compounding all of these problems is the failure of the classical PBT approach to consider, in any way, the quantity of the material released to the environment, a parameter that is critical to exposure assessment (Mackay et al. 2003b).

Metals are obviously persistent in the sense that they do not degrade to CO_2, water, and other elements. The conventional concept of persistence as developed for organic chemicals cannot, however, be satisfactorily applied to metals (Skeaff et al. 2002). Metals usually exist as several species that can undergo reversible or irreversible interconversion among, for example, dissolved species and sparingly soluble salts. All metals have natural background concentrations established by local biogeochemical processes, and some of the metals are essential micronutrients. The uptake by, and release of, metals from organisms may be modulated by physiological processes and exposure conditions (for example, acclimation). Organisms differ widely in their tolerance to metals, with some organisms being able to store certain metals with no adverse physiological response (Mason and Jenkins 1995).

3.1.2 A Unifying Model

Targeted efforts have been put forth in an attempt to fit metals into the PBT paradigm (Adams et al. 2000; Di Toro and Paquin 2000; McGeer et al. 2003; Existing Substances Branch 2003; Mackay et al. 2003a). For example, analyses of the degree of partitioning of a variety of metals have been performed to provide insight concerning their persistence in the water column and the rate of delivery of sorbed metal to aquatic sediments, that is, to transfer the risk from the water column to the sediments. Other types of analyses have included the evaluation of metal speciation in the water column as a way to consider metal bioavailability and the development of models (for example, quantitative structure activity relationships [QSARs]) to more fully characterize the potential for bioaccumulation and toxicity. Although these types of analyses had the potential to help broaden the scientific underpinnings of the PBT analysis, the difficulty of prescribing a meaningful way to quantitatively weight the various PBT parameters and to integrate them into a single numerical value suitable for use in a ranking analysis remained as unresolved problems. Such limitations may be overcome by integrating disparate PBT analyses, through use of a suite of evolving computational modeling tools, into the UWM.

This new approach reflects a different way of predicting, or assessing, the environmental fate and effects of chemicals. Such an approach preserves the utility of the supporting data that are called for in the context of current regulatory procedures. It also continues to consider PBT, though it does so in a less direct but more holistic evaluation framework. The UWM concept embodies the development of a methodology for evaluating both metals and, eventually, organic chemicals in a unified framework in which decisions are based on a more environmentally meaningful simulation of fate processes than is presently the case, incorporating the current state of science for both chemical classes. This chapter illustrates the principles underlying the proposed UWM approach, identifies the nature of the data required, and demonstrates the kind of results and output that will be generated.

The UWM, as it is proposed here, is a conceptual model that is envisaged for use in the hazard assessment and priority ranking of metals and metal compounds for their environmental effects. The data needs for such a holistic model are clearly significant, and it must be accepted that at the present time, for many metals there is insufficient information available to adequately assess them. Even in these cases, however, the UWM may provide a conceptual framework that can guide future data gathering.

In this chapter, the focus is on hazard assessment, specifically the ranking of the potential deleterious effects of metals within a single, standardized conceptual ecosystem. In developing and discussing the UWM, it is important to consider as completely as possible all the significant interactions that affect metal behavior in the environment in order to gauge their relevance to hazard assessment. Risk assessment might also be done through the UWM approach, but that would require local conditions to be taken into account, and a series of site-specific UWMs would be required.

3.2 THE UNIT WORLD MODEL (UWM)

Because the PBT approach does not reflect all of the important processes controlling fate for either organic chemicals or metals, it can result in inconsistencies in the evaluations that are performed for both types of substances. It is thus necessary to consider a more comprehensive approach — one that considers a more complete suite of fate processes. One possible solution is to integrate PBT into a more comprehensive model framework. With this approach, a relatively simple screening-level hazard assessment can be performed that accommodates metal-specific characteristics such as speciation and sensitivity to redox conditions, while at the same time also being applicable to organic substances. For example, the distribution of metals among phases is governed by numerous chemical reactions and biological processes rather than the simple equilibrium partitioning approach that is often a good approximation for organics. Consequently, metal partitioning can be nonlinear, and metal chemistry and fate are highly dependent on the chemical and biological characteristics of the ambient environment. Persistence (e.g., for organics) or residence time (e.g., for metals) is still considered in this framework, as are uptake, toxicity, and other processes controlling fate that also have a bearing on fate and exposure of each of these groups of substances. For example, particulate and diffusive transport is also appropriately reflected in the evaluation. Use of the UWM approach, which is still to be evaluated, may satisfy the need to subject all chemicals in commerce, including metals, to a consistent, transparent, and equitable assessment system. The advantages of this approach include:

- Avoidance of contentious and nonproductive debate about the PBT properties of metals.
- Retention of a consistent system for evaluating metals and organics, which should permit direct comparison of hazard for these classes of substances.
- Fidelity to characteristic properties and mechanisms governing the distribution and fate of both substances in the environment.
- Realistic and appropriate categorization and hazard and screening assessments that enable protection of the environment.

The UWM, as applied to metals, is predicated on toxicity evaluated through the same modeling framework that has been successfully used in other regulatory applications. Additional details regarding toxicity data for application in the UWM are provided in Chapter 4 (bioaccumulation), Chapter 5 (toxicity), and Chapter 6 (terrestrial). The UWM is based on models derived from previous modeling efforts for metals for aquatic systems (e.g., Di Toro 2001; Bhavsar et al. 2004a, 2004b). These models range from the highly sophisticated to the relatively simple; some have been evaluated with field data.

For hazard assessment, the UWM would be run for a generic environment, giving output in the form of substance-specific loadings or concentrations that would result in accumulations in target compartments that equal specified toxicity thresholds, termed "critical limits" (for example, LC_{50}s [lethal concentration to 50% of test organisms], EC_{50}s [effective concentration to 50% of test organisms], NOECs [no-observed-effect concentrations], or PNECs [predicted no-effect concentrations]). Such UWM loadings or concentrations may be ranked in order from lowest (representing the greatest hazard) to highest (representing the least hazard). It may be possible to use such outputs in both classification and priority ranking. The model could, in principle, also be used for regional screening assessments, that is, risk assessments, but that would require significant additional model developmental work beyond that envisaged here to achieve this objective.

The loading approach proposed here follows methodologies already being applied or developed for effects-based risk assessments of acid deposition and metals (Doyle et al. 2003). In such contexts, the term *critical load* is used to denote the steady-state loading, which results in the system reaching a critical limit for environmental damage. Different terms, for example, target load, may be used if time dependence is considered. The present proposal, at least initially, is to calculate steady-state loads for one or more generic environments j, and these loads are therefore referred to as CL_j. It should, however, be noted that the concept of critical or target loading is not currently accepted in many countries as a criterion to be used in setting environmental guidelines.

In quantitative terms, an evaluative multimedia model provides for a given emission rate E (mol/h or g/h) that results in a corresponding critical concentration in water, C_W (mol/l or g/l), and sediment, C_S (mol/kg). For metals, C_W and C_S can refer to any particular form present. By running the model for evaluative conditions, the critical value of E can be sought, that is, E_C, which will yield a value of C_W equal to the LC_{50} (or some other set of alternative regulatory effect levels that are used for purposes of the ranking analysis). This value of E is the critical load to the system.

When this approach is used, metals can thus be ranked in terms of environmental hazard by comparing their critical E values as a critical load for a defined system or set of systems. An advantage of this approach is that partitioning, transport, and toxicity information are integrated into a mechanistic model even if the data are not available to evaluate the model. Further, the method is not limited to metals, as a critical load can be calculated analogously for organic substances as well.

The implementation of such an approach requires the following:

- The number, nature, and properties of the relevant compartments.
- Representative intermedia transport parameters such as soil runoff and sediment deposition rates.
- Clear understanding of the chemical and biological behavior of the metal in each compartment.

- Relevant subroutines on bioaccumulation and toxicity in all model compartments.
- Operation of the model in steady-state or dynamic modes.
- Mode of introduction of loadings to the generic environment (unit world), for example, to water, or soil, or both directly or by atmospheric deposition.
- Uncertainty analysis.
- An evaluation program.

3.3 HAZARD ASSESSMENT FRAMEWORK FOR A GENERIC ENVIRONMENT

The model proposed is designed to balance the competing needs of simplicity and transparency on the one hand, and realism on the other. It is a mass balance model that describes the fate of metals in a generic setting comprising both terrestrial and aquatic compartments. Models of this sort are commonly used for organic chemical priority ranking and screening-level risk analysis (Mackay et al. 1985, 1992, 1996). The UWM is similar to these existing models, but includes the processes that are necessary to describe the behavior of metals in aquatic and terrestrial environments. These processes include speciation, complexation, precipitation, and transformation/dissolution of metals in the soil, water column, and sediment compartments. The UWM is not intended to represent a specific location, but rather a representative setting (a watershed including all environmental compartments — air, water, and soil) that is typical of the class of environments being evaluated. It is also not intended to be a complete description of metal fate and transport. Rather, it focuses on the primary processes that affect the long-term fate and toxicity of metals. It is designed to be used for evaluative purposes, rather than for detailed site-specific assessment.

3.3.1 GENERALIZED MODEL FRAMEWORK

It is unwise at this stage to define a specific UWM and expect that it will stand the test of time. Rather, we suggest a general structure of a model in the full expectation that it will change in the light of experience. It may be that for some evaluations, only an aquatic system need be considered but, for others, a terrestrial system will be necessary. Further, the optimal degree of vertical segmentation in soil and sediment is not yet established. Different modelers favor different approaches; thus, it is hoped that this proposed methodology will encourage a diversity of approaches in an open and constructively competitive atmosphere.

The conceptual model framework is presented in Figure 3.1. It is composed of aquatic and terrestrial sectors. These are divided into completely mixed volumes that represent the various model compartments. The principles underlying the construction of these types of models are well understood and detailed descriptions are available (Thomann and Mueller 1987; Schnoor 1996; Chapra 1997); in addition, aspects of the models have been developed previously for general and specific applications (e.g., Diamond 1995; Diamond et al. 2000; Mayer et al. 2002; Bhavsar et al. 2004a). These are fate and behavior models that do not encompass all the

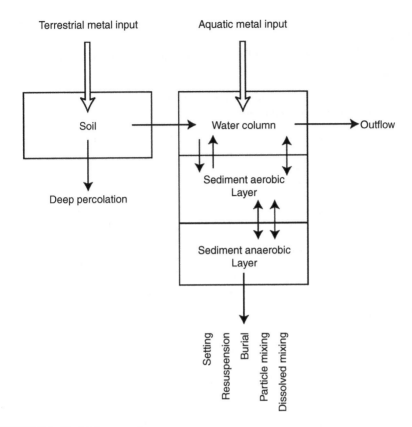

FIGURE 3.1 Model framework.

elements of the hazard or risk assessment process, but may provide vital components of a UWM.

Mass balance equations can be written, for each of the 4 compartments (soil, water column, aerobic sediment, and anaerobic sediment [Figure 3.1]), either as algebraic equations describing the steady-state system, or as differential equations describing the unsteady-state, or dynamic system. Emissions are defined. These equations contain 4 unknowns representing the quantity or concentration of substance in each compartment. The equations can be solved to yield concentrations, magnitude of the masses, and fluxes, including reaction rates. These results depend on both the emission and the mode of entry, that is, whether the emission is to air, water, soil, or a combination of these 3 compartments. An overall persistence or residence time can be calculated as the ratio of the total quantity of chemical (kg) present at steady state to the rate of loss (kg/h).

3.3.2 Water Column/Sediment Model

The water column/sediment model is illustrated on the right-hand side of Figure 3.1 and in Table 3.1. Models of this sort are described in detail elsewhere (Di Toro

TABLE 3.1
Equations of Water Column/Sediment Model for the Water Column/Sediment Compartments

Mass Balance Equations

Water column — total metal concentration

$$H_0 \frac{dC_{T0}}{dt} = J_T - w_{01} f_{p0} C_{T0} + w_{10} f_{p1} C_{T1} - K_{L01}(f_{d0} C_{T0} - f_{d1} C_{T1}) + k_{S,Ox,0} C_{S0} H_0 \\ + F_{Soil-Water} - F_{Outflow}$$

Aerobic sediment layer — total metal concentration

$$H_1 \frac{dC_{T1}}{dt} = \begin{aligned} & w_{01} f_{p0} C_{T0} - w_{10} f_{p1} C_{T1} - w_2 f_{p1} C_{T1} - K_{L01}(f_{d1} C_{T1} - f_{d0} C_{T0}) \\ & - K_{L12}(f_{d1} C_{T1} - f_{d2} C_{T2}) - w_{12}(f_{p1} C_{T1} - f_{p2} C_{T2}) + k_{S,Ox1} C_{S1} H_1 \end{aligned}$$

Anaerobic sediment layer — total metal concentration

$$H_2 \frac{dC_{T2}}{dt} = \begin{aligned} & w_{10} f_{p1} C_{T1} + w_2(f_{p1} C_{T1} - f_{p2} C_{T2}) - K_{L12}(f_{d2} C_{T2} - f_{d1} C_{T1}) \\ & - w_{12}(f_{p2} C_{T2} - f_{p1} C_{T1}) - k_{S,P,2} f_{d2} C_{T1} H_2 \end{aligned}$$

Water column — metal sulfide concentration

$$H_0 \frac{dC_{S0}}{dt} = -w_{01} C_{S0} + w_{10} C_{S1} - k_{S,Ox1} C_{S0} H_0$$

Aerobic sediment layer — metal sulfide concentration

$$H_1 \frac{dC_{S1}}{dt} = w_{01} C_{S0} - w_{10} C_{S1} - w_2 C_{S1} - w_{12}(C_{S1} - C_{S0}) - k_{S,Ox1} C_{S1} H_1$$

Anaerobic sediment layer — metal sulfide concentration

$$H_2 \frac{dC_{S2}}{dt} = +w_{10} C_{S1} + w_2(C_{S1} - C_{S2}) - w_{12}(C_{S2} - C_{S1}) + k_{S,P,2} f_{d2} C_{T2} H_2$$

Definitions

Description	Symbol	Units
Water-column depth	H_0	(m)
Aerobic sediment layer depth	H_1	(m)
Anaerobic sediment layer depth	H_2	(m)
Total (particulate + dissolved) metal concentration in layer j	C_{Tj}	(mmol/l)
Areal loading rate to the water column	J_T	(mmol/m²d)
Particulate fraction in layer j	f_{pj}	unitless
Dissolved fraction in layer j	f_{dj}	unitless
Particle settling velocity from the water column to the aerobic sediment layer	W_{01}	(m/d)
Particle resuspension velocity from the aerobic sediment layer to the water column	W_{10}	(m/d)
Particle mixing velocity between the aerobic and anaerobic sediment layer	W_{12}	(m/d)
Burial velocity (= sedimentation velocity)	W_2	(m/d)

Continued.

TABLE 3.1 *(Continued)*
Equations of Water Column/Sediment Model for the Water Column/Sediment Compartments

Diffusive mass transfer coefficient between water column and aerobic sediment layer pore water	K_{L01}	(m/d)
Diffusive mass transfer coefficient between aerobic and anaerobic sediment layer pore water	K_{L12}	(m/d)
Oxidation rate of metal sulfide in layer j	$k_{S,Ox,j}$	d^1
Precipitation rate of metal sulfide in layer 2	$k_{S,Pj}$	d^1

Process	Assumption
Physical representation — soil	Two vertically connected well-mixed boxes, each with an outflow to a water body; percolation to groundwater
Physical representation — aquatic	Well-mixed oxic water overlying two well-mixed sediment compartments, oxic and anoxic
Input to lake and soil	The dissolved fraction of the added substance is a constant ratio of the total quantity added
Solution speciation and partitioning to DOM in soil/waters/sediment	Can be calculated using a geochemical model such as WHAM6
Partitioning to nonsulfidic particles in water and sediment	Can be predicted by SCAMP assuming only organic matter and Fe and Mn oxides are responsible for binding, and they are present in fixed fractions
Particle formation/transport	Particles settle from water at a constant rate and are immediately replenished, maintaining a constant concentration
Particle aging	A fraction of metals in soils is removed by irreversible binding to a particle
Transfer of solutes between water/upper sediment/lower sediment	Driven by concentration gradients (diffusion) at interfaces
Resuspension from sediment to water	Occurs as a continuous transfer of particles
Bioturbation	Occurs as a continuous exchange of particulate material between sediment layers
Burial	Occurs as a loss from the lower layer by fixing the total volume of sediment
Oxide reduction	Occurs instantaneously in oxic sediment if critical set Eh is met
Sulfate reduction	Occurs in anoxic sediment at a specified rate; all metals are bound as sulfides; metals in excess of sulfide partition to any remaining oxides and POC
Sulfide oxidation	Occurs through bioturbational transfer at a fixed rate
Toxicity	Determined potentially in four compartments by: (1) total concentration in the surface soil; (2) concentration of components in the aqueous phase in waters; (3) concentration of components in the aqueous phase in oxic sediment, assuming slow biological uptake processes; and (4) concentration in oxic sediments as the solid phase that might be ingested

2001), so only a brief description is given here. Two state variables are modeled in the 3 compartments: the total dissolved and sorbed metal concentration (C_T) and the concentration of metal sulfide (C_S). The water-column compartment is assumed to be completely mixed and oxic. It represents a well-mixed shallow lake or reservoir in which water inflow and outflow are neglected. The water column and sediment pore water interact via diffusion of dissolved metal species augmented by bioirrigation. Dissolved metal partitions onto particles in the water column that then settle into the sediment. The sediment is modeled as 2 layers: an aerobic layer in which the oxygen concentration is greater than 0, and an anaerobic layer, representing the zone of sulfate reduction. This minimum representation is necessary because of the importance of redox variation and sulfide formation on metal fate and toxicity. In addition to metal removal from the water column through particle settling, particles and associated metal are resuspended from the aerobic layer to the overlying water. Particles and particle-sorbed metal are also mixed between the aerobic and anaerobic layers by bioturbation, and pore water mixes because of diffusion and bioirrigation. Finally, particles and their associated metal are removed by burial.

For the modeling of fate processes, it is necessary to specify the fractions of the metal that are in the dissolved and particulate phases because they are transported by different processes, for example, particle settling transports only particle-bound metal to the sediment. An empirical partition coefficient would suffice for this purpose. However, because this model is being designed to apply to many metals and metal compounds, it is preferable to have a consistent method for computing partitioning. Several speciation–complexation models have been developed, which estimate metal speciation in the aqueous phase, and complexation to a solid phase, assuming equilibrium conditions. In this example, partitioning in the water column and aerobic sediment layer may be computed using chemical speciation models such as WHAM (Windemere Humic Aqueous Model) 6/SCAMP (Tipping 1998; Lofts and Tipping 1998). These models have been calibrated with laboratory data and have parameters for many, but not all metals. Some field testing has also been performed with reasonable results (Lofts and Tipping 1998; Bryan et al. 2002). Aqueous phase speciation includes dissolved organic carbon (DOC) complexation. The particulate partitioning phases are organic carbon, Mn and Fe oxides, and a mineral cation exchanger. The concentrations of these particulate phases are specified externally as part of the input parameters. SCAMP assumes that the partitioning to these phases is additive.

The importance of metal sulfide precipitation in the anaerobic layer and subsequent oxidation in the aerobic layer is well known, and models of these phenomena have been developed (Boudreau 1991; Di Toro et al. 1996). Therefore, these reactions are modeled explicitly. Metal sulfide precipitate is formed until the sediment sulfide is exhausted. Metal partitioning to particulate organic carbon is included if the available sulfide is exhausted. Therefore, the pore water metal concentration is effectively 0 in the presence of excess sulfide, or determined by organic carbon partitioning using a chemical speciation model, such as WHAM6.

The model is formulated as a series of mass balance equations that are listed in Table 3.1. The equations are formulated assuming that the rates of adsorption and desorption are fast relative to other processes. This is the local equilibrium assumption. By contrast, the kinetics of metal sulfide precipitation and dissolution are formulated as kinetic processes. The concentrations and characteristics of the necessary water column and particulate partitioning phases are established to represent the generic environments to be used in the evaluation.

3.3.3 Soil Model

The soil model comprises a single mixed box, containing solids and solution, as shown on the left-hand side of Figure 3.1. The soil receives the metal of interest in the soluble form. Physical and chemical conditions are specified. For a whole-catchment model, drainage of the soil solution would contribute to the surface waters.

Within the United Nations Economic Commission for Europe/Convention on Long-Range Transboundary Air Pollution (UNECE/CLRTAP), the Expert Group on Heavy Metals have developed methods for calculating critical loads of metals to different terrestrial ecosystems, that is, a risk-based assessment. Steady-state conditions are considered, and the critical load is that corresponding to the critical limit, a concentration of metal that is the maximum allowable, in respect of ecosystem damage. As discussed by DeVries and Bakker (1998), there are several fluxes that govern steady-state metal concentrations in soil, the principal ones being:

F_{in} — input flux of (reactive) metal from external sources
F_{weath} — weathering input
F_{age} — removal by aging processes in mineral phases
F_{ppt} — removal by the formation of precipitates
$F_{harvest}$ — removal in harvested plants
F_{vola} — removal by volatilization
F_{dust} — removal in wind-blown dust
F_{drain} — removal in drainage water

At steady-state, the fluxes balance as follows:

$$F_{in} + F_{weath} = F_{age} + F_{ppt} + F_{harvest} + F_{volat} + F_{dust} + F_{drain} \qquad (3.1)$$

To a first approximation, all the fluxes on the right-hand side of Equation 3.1 depend on the amount of metal in the system, whereas F_{weath} can be assumed to be independent. The right-hand side terms can, in principle, be calculated if the distribution of metal between the solid and aqueous phases is known, and if the speciation of metal in the soil water is known.

The total metal concentration in soil water, $[M]_{SW}$, is given by:

$$[M]_{SW} = [M_{FI}] + [M_{inorg}] + [M\text{-}DOM] + [M\text{-}SPM] \qquad (3.2)$$

where M_{FI} is the free ion (e.g., Zn^{2+}, AsO_4^{3-}), and M_{inorg}, M-DOM, and M-SPM are metal present in inorganic complexes, bound to dissolved organic matter (DOM), and sorbed to mobile suspended particulate matter (SPM), respectively. The total reactive metal sorbed to the soil solids (e.g., Q_M in mol/g^{-1}) can be expressed as:

$$Q_M = K_D [M]_{SW} \qquad (3.3)$$

or

$$Q_M = K_D^* [M_{FI}] \qquad (3.4)$$

or by more complex expressions involving more solution species. Note that K_D and K_D^* will depend upon solution conditions and on the composition of the soil solids. Given appropriate modeling capabilities and ancillary data, Equation 3.2 through Equation 3.4 describe the entire speciation of the metal in the soil–water system. Given appropriate models and data, the fluxes on the right-hand side of Equation 3.1 can be determined. For example, in systems where there is an excess of precipitation over evaporation, that is, where there is a net downward loss of water from the soil, F_{drain} can be expressed as the product of the steady-state concentration of metal in soil water, and the water drainage flux:

$$F_{drain} = [M]_{SW} \cdot w_F \qquad (3.5)$$

The key remaining issue is to find the critical concentration of metal, that is, a measure of the maximum permissible toxic effect.

The conventional means of expressing toxic effects in soils is by the total reactive soil metal (e.g., mol g^{-1}). In most soils *in situ*, this will be equivalent to the total metal sorbed to the soil solids, because the amount of metal in the soil water will be only a small fraction of the total. Thus, we can equate the total reactive soil metal with $Q_{M,tox}$, calculate the concentrations of the aqueous metal species, derive the removal fluxes in Equation 3.1 and, given F_{weath}, thereby obtain the critical load, $F_{in,tox}$. The smaller $F_{in,tox}$, the more toxic the metal.

An alternative means of expressing metal toxicity in soils is to use the free metal ion concentration. This is mechanistically reasonable for soil organisms that are exposed to metal through the solution phase, rather than from the solid, although there may be cases where it should not be applied. As recognized in the Biotic Ligand Model (BLM), the free ion concentration alone is insufficient to characterize toxicity in general, and the interfering effects of other solutes should be accounted for. If we assume for the moment that a free-ion toxicity criterion could be used, then we can obtain $[M]_{SW}$ from Equation 3.2, given the ability to compute $[M_{inorg}]$, [M-DOM], and [M-SPM] from $[M^{z+}]$. In this way, we can find F_{in} without taking any account of the interaction of the metal with the soil solids; at steady state, this has no influence on the solution composition. For 4 metals (Cu, Zn, Cd, and Pb), soil toxicity data have been analyzed and pH-dependent free-ion critical limits derived (Lofts et al. 2004). For risk assessment purposes, the free metal ion approach is preferable to characterizing toxicity in terms of the total reactive soil metal, because it allows variations in solution chemistry to be taken into account, and is

TABLE 3.2
Key Processes of the UWM

	Soil (Terrestrial Environment)	Water	Sediment
Ecotoxicity reference level	X	X	X
Bioaccumulation	X	X	X
Partitioning	X	X	X
Speciation	X	X	X
Aging	X		
AVS binding			X
Sediment burial			X

independent of the soil solids composition. However, for hazard assessment, this is less important (see Section 3.6.2).

3.3.4 KEY PROCESSES

There are a few key processes that significantly influence the hazard ranking or absolute critical load outcome of the model, as shown in Table 3.2. These are the ecotoxicity and bioaccumulation, the speciation and partitioning reactions among water, sediment, and soil, and the role of the biota that determine both fate (the compartments where the metal finally resides) and bioavailability. Additionally, the transformation of metals into nonbioavailable forms has a significant impact on the critical load. In this model, other transformation processes that are not well understood include aging reactions in soils, acid volatile sulfide (AVS)-binding of the metal, and deep burial of metal in sediments (i.e., burial such that the metal is effectively sequestered and prevented from interaction with the overlying biotic compartment). The properties of a metal affecting the key processes are, therefore, those properties that need to be the most accurately estimated for correct hazard identification. Additionally, the properties of the system such as surface, soil, and pore water chemistry, strongly influence the outcome of the model — the critical load as well as relative ranking of metals with respect to one another (Bhavsar et al., in preparation). As such, system characterization is an important component of developing the UWM. Note that biological interactions such as predation can also be key processes, but are not included in the UWM at this time.

3.4 SOURCE TERM

There are potentially 3 critical issues that are pertinent to the source term for metals, but that are not normally considered for organic compounds: (1) how to treat the natural occurrence of metals in the environment, (2) how to translate an amount of metal in commerce into a loading of the generic watershed environment, and (3) how to combine the various compounds used in commerce into a loading function.

3.4.1 NATURAL OCCURRENCE OF METALS

Metals are natural constituents of all rocks, soil, water, air, and sediments, which over geological time have attained a distribution in various environmental compartments. Having evolved in the presence of metals, all life-forms are therefore exposed to low concentrations of metals that occur naturally in their habitat (Nriagu 1990). In ecosystems, some metals serve as essential nutrients, whereas others are toxic and have no known beneficial biological function. At a high enough dose all metals are, however, biologically toxic.

For environmental risk assessment, metals from natural sources are indistinguishable from those from commercial sources in terms of their physicochemical properties, bioavailability, and toxicity. The naturally occurring metals can carry as much risk as metals from commercial sources and must be included in site- or region-specific risk assessment (Struijs et al. 1997; Crommentuijn et al. 2000). However, for hazard assessment, that is, classification and ranking of potential dangers from anthropogenic materials, it could be advantageous to not include background metals in the unit world. This makes the assessment much clearer, and avoids the introduction of arbitrary "representative" weathering rates. However, it also makes the model more conservative and less realistic.

3.4.2 DETERMINING THE INPUT TERM

The model, as presently constructed, requires the input of the metal in the aqueous species that is most likely to be present under average environmental conditions. However, metal-containing substances in commerce are often sparingly soluble (metal sulfides and oxides as powders or massives), and differ widely in dissolution rates and in the dependence of dissolution rate on ambient conditions. Thus, it is necessary to have a method to provide a representative value of the fraction of the sparingly soluble substance that might be expected to enter the environment in soluble form. In contrast, for organic compounds, it is typically assumed that the natural background level is 0 and that the fraction of the compound available to enter the environment is 100%.

For assessing the hazard of metal substances, the model requires specification of the loading to the system. For organic substances, this loading term is derived as the compound itself. For metals and alloys in commerce, determining the loading term is more complex. The model, as presently constructed, requires the input of metal in dissolved form. However, determining this fraction of total metal is difficult because the forms in commerce differ widely in rates of reaction with aqueous media and in the dependence of these rates on ambient conditions.

A few approaches are now used to measure dissolution rates under specified, standardized conditions (Skeaff et al. 2000; OECD 2001b; UN 2003). These approaches result in an operationally defined dissolution term for hazard assessment purposes. Blowes et al. (1999, 2003) have estimated dissolution rates of specific metals from complex systems under ambient conditions. It is clear from a review of these approaches that many uncertainties remain in determining representative and consistent dissolution rates that are compatible with the needs of the UWM.

Integrated Assessment of Hazards for Metals and Inorganic Metal Compounds
LOADING TERM Assessment

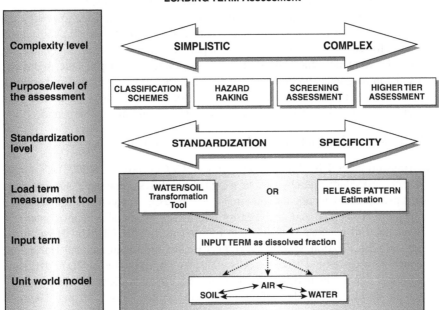

FIGURE 3.2 Aims and objectives of the hazard assessment.

The approach recommended here for determining the input term for the model is consistent with the different aims and objectives of hazard assessment (Figure 3.2). First, the critical emission rate $E_{c(soluble)}$ is determined based on the introduction of the metal in a soluble form. Next, a transfer factor (the fraction of the sparingly soluble parent compound that is rendered soluble in standard dissolution tests — Skeaff et al. 2000; OECD 2001b) is determined. Finally, the critical emission rate based on introduction of the metal in soluble form $E_{c(soluble)}$ is divided by the transfer factor to yield the effective critical emission rate of total metal for assessment purposes. For example, if the critical emission rate based on introduction of the soluble form is 100 mol/h, and the transfer factor is 0.01 $mol_{soluble}/mol_{parent\ compound}$, the effective value of E_c for assessment purposes is 10,000 mol/h.

This methodology is a highly simplified approach designed for a hazard assessment in the context described above. For a more realistic assessment, an approach that respects the dependence of this transfer factor on specific environmental conditions would be appropriate. Other approaches can and have been used for hazard assessment. For example, for the purposes of aquatic hazard ranking, a simple standard test of a substance introduced into an aqueous medium to determine an initial estimate of total dissolved metal released has been used (OECD 2001b). Such a test is not necessarily entirely founded in environmental realism; it has only the limited applicability for which it was intended and designed. Under the appropriate conditions, such a test is able to distinguish between a metal and its readily and sparingly soluble compounds, and rank them based on transfer factor. Similarly, the test would also be able to rank other metals and metal compounds.

For the purposes of higher tier assessments, capability is needed to predict the actual and eventual fate of metals released to and the effect of such exposure on species in the environment. It is clear that sophisticated research, experimental tests, and predictive models encompassing metal distributions among aquatic, sediment, and soil compartments, and interactions therein, need to be developed and applied. In the longer term it is hoped that this aspect of the model may be able to take account of the dissolution kinetics of different metals and metal compounds.

3.4.2.1 Measuring Tool for the Aquatic Compartment

The Transformation/Dissolution (T/D) Protocol (OECD 2001b; UN 2003) is a simple laboratory procedure whereby various weighed quantities of the metal substance are loaded to a standard aquatic medium in a reaction vessel, subjected to agitation, and the medium is sampled and analyzed for total dissolved metal at regular intervals. It is intended to generate data that can be used in the UN Globally Harmonized Integrated Classification System for Human Health and Environmental Hazards of Chemical Substances and Mixtures (UN 2003) developed by the OECD (2001a) and to represent conditions normally found in the environment. Data for acute classification are developed in 7-day T/D tests of substance loadings 1, 10, and 100 mg/l in an aqueous medium at a pH in the range 6 to 8 that should maximize the dissolution of the substance, usually at pH 6. For chronic classifications, the current OECD (2001b) T/D Protocol also provides for a 28-day test of a 1 mg/l loading. To establish the classification of the substance, the value of $Me_T(aq)$ (representing the dissolved metal) at 7 days is compared to an $L(E)C_{50}$ selected for the classification of the metal substance. The mass loading that delivers the $L(E)C_{50}$ to the medium determines the acute hazard classification of the substance. Although this test was developed for a different purpose, it could be adapted to estimate transfer factors, as defined in Section 3.4.2.

3.4.2.2 Measuring Tool for the Soil Compartment

It is also desirable to assess the rate of release of bioavailable forms of a metal from a parent compound in the soil compartment. There are a number of dissolution pathways that lead to the release of free metal ions from metal-containing substances, including direct pathways, or indirect oxidative or reductive pathways. Sulfides disposed subaerially have been observed to dissolve more slowly than in aerial settings (Pedersen 1993). Biological catalysis of mineral dissolution is well documented, and similar catalysis for many commercial forms of metals is expected (Nordstrom and Southam 1997). Sorption of metal in soil decreases free metal activity in solution, thereby creating a sink for metal dissolution. These differences potentially lead to reaction pathways different from those in the aquatic setting. Examples include: the behavior of elemental Fe (Blowes et al. 1999), the disposal or dispersal of metal sulfide ores (Blowes et al. 2003), the stability of metal sulfate salts at mine sites (Nordstrom and Alpers 1999; Frau 2000), and the stability of nitrate salts at fertilizer plants.

TABLE 3.3
Generic Data Needed to Feed the Model

	Surface Soil	Subsurface Soil	Water Column	Aerobic Sediment	Anaerobic Sediment
Fraction soluble		-x-		-x-	-x-
Metal loading rate		-x-		-x-	-x-
Inflow/outflow		-x-		-x-	-x-
Surface area, A					
Depth, D	-x-	-x-			
Outflow fraction			-x-	-x-	-x-
Solids concentration					
Soil aging term			-x-	-x-	-x-
Solid density					
Porosity	-x-[a]	-x-	-x-		
Fraction$_{oc}$	-x-	-x-			-x-
Fraction$_{Mn}$	-x-	-x-			-x-
Fraction$_{Fe}$	-x-	-x-			-x-
Settling rate	-x-	-x-		-x-	-x-
Resuspension rate	-x-	-x-	-x-		-x-
Burial rate	-x-	-x-	-x-	-x-	
Particle mixing	-x-	-x-	-x-		
Diffusion coefficient	-x-	-x-	-x-		
Transfer function			-x-	-x-	-x-
pH					-x-
pCO$_2$				-x-	-x--
Major ions					-x-
Background metal weathering		-x-		-x-	-x-
rates		-x-	-x-	-x-	-x-
Dissolved organic carbon					-x-
Eh	-x-	-x-	-x-		-x-
Sulfide precipitation rate	-x-	-x-	-x-		-x-
Sulfide oxidation rate	-x-	-x-	-x-		-x-
Reference toxicity values		-x-			-x-
Henry's constant[a]		-x-	-x-		-x-
Biodegradation K[a]					
Diffusivities		-x-			-x-
Temperature					

Note: A general terminology that can be applied to all compartments is used. Crosses indicate that they are not applicable to that compartment.

[a] Organic chemicals.

Dissolution rates of sparingly soluble metal substances will depend on the composition and particle sizes, the surface area, surface roughness factors, chemical composition of the pore water, and the overall manner in which the material is disposed or dispersed including dissolution or weathering rates. Materials dispersed will react more rapidly than materials stored in piles; standard tests have been developed to evaluate the variability in reactivity of the various sulfide minerals (Blowes et al. 2003). These tests have been conducted in a number of ways, ranging from simple mixed reactor dissolution tests to dynamic flow tests that provide variable moisture contents to emulate wetting and drying cycles. The composition of the medium and length of tests is therefore critical. Matrices representative of the soil environment are required, including pH values, redox conditions, and organic matter contents representative of given soil environments. If acids or bases are produced during the dissolution process, consideration of the influence of these releases on dissolution is required. Testing per Blowes et al. (2003) for a representative soil or soils should assess: (1) the rate of metal release from the pure compound, and (2) the further interaction of the released metal with the soil or soil pore water.

3.4.3 Combinations of Commercial Compounds

Whereas for risk assessment, different combinations of source terms would have to be considered, hazard assessment is best applied to single sources. For example, a soluble metal salt would be hazard-assessed differently from an ingot of the same metal.

3.4.4 Generic Data Needs

Application of the UWM requires considerable information on the generic properties of the systems. Table 3.4 summarizes the parameters required for each compartment. When parameterizing the UWM, it is essential to recall that its use is for generic hazard assessment, not risk assessment. As such, the UWM must be parameterized with data that are representative of a range of systems while not reproducing any particular system. The same generality applies to seasonal and climatic variations.

As with any model, the choice of parameter values strongly influences model estimates. For example, Bhavsar et al. (in preparation) explore the importance of parameterization in ranking metals using TRANSPEC, a loosely coupled metal speciation–complexation and fate model developed for metals. They found that choice of parameter values in a lake model can significantly alter the ratio of loading to resultant free metal ion concentration. Although the ranking of metals did not change among the scenarios examined, the magnitude of the metal-to-metal differences did change considerably as a function of ambient chemistry and lake characteristics.

3.5 APPLICATION OF THE UWM

The UWM can be applied to classification, ranking, and screening using a tiered system outlined in the following sections.

TABLE 3.4
Simple Water-Column Sediment Model:
Equations, Parameters, and Solutions

Water Column — Sediment 2-Layer Model — Conservative

Water Column (1)

$$H_1 \frac{dC_{T1}}{dt} = -w_1 f_{p1} C_{T1} - w_{12}(f_{p1} C_{T1} - f_{p2} C_{T2}) - K_{L12}(f_{d1} C_{T1} - f_{d2} C_{T2})$$

Sediment (2)

$$H_2 \frac{dC_{T2}}{dt} = +w_1 f_{p1} C_{T1} + w_{12}(f_{p1} C_{T1} - f_{p2} C_{T2}) + K_{L12}(f_{d1} C_{T1} - f_{d2} C_{T2})$$

Solutions

$$C_{T1}(t) = C_{T1}(0)\left(\frac{s_1}{s_1 + s_2} \exp(-(s_1 + s_2)t) + \frac{s_2}{s_1 + s_2} \right)$$

where

$$j_{1\rightarrow2} = w_1 f_{p1} + w_{12} f_{p1} + K_{L12} f_{d1}$$

Flux to sediment

$$s_1 = j_{1\rightarrow2} / H_1$$

Flush out time in water column

$$j_{2\rightarrow1} = w_{12} f_{p2} + K_{L12} f_{d2}$$

Flux to water column

$$s_2 = j_{2\rightarrow1} / H_2$$

Flush out time in the sediment

Water Column Concentration

$$C_{T1}(t) = C_{T1}(0)\left(\underbrace{\frac{s_1}{s_1 + s_2} \exp(-(s_1 + s_2)t)}_{\text{Fraction removed to sediment}} + \underbrace{\frac{s_2}{s_1 + s_2}}_{\text{Fraction remaining in water column}} \right)$$

$$\frac{s_1}{s_1 + s_2} = \frac{j_{1\rightarrow2}}{j_{1\rightarrow2} + j_{2\rightarrow1}\left(\dfrac{H_1}{H_2}\right)} = \frac{1}{1 + \left(\dfrac{j_{2\rightarrow1}}{j_{1\rightarrow2}}\right)\left(\dfrac{H_1}{H_2}\right)}$$

$$\left(\frac{j_{2\rightarrow1}}{j_{1\rightarrow2}}\right)\left(\frac{H_1}{H_2}\right) \rightarrow 0 \text{ (goes to sediment)}$$

Continued.

TABLE 3.4 *(Continued)*
Simple Water-Column Sediment Model:
Equations, Parameters, and Solutions

Orders of magnitude

$$w_1 = 1 \text{ (m/d)} = 10^0 \text{ (m/d)}$$
$$w_{12} = 3 \text{ (cm/yr)} = 10^{-4} \text{ (m/d)}$$
$$K_{L12} = 10 \text{ (cm/d)} = 10^{-2} \text{ (m/d)}$$
$$H_1 = 10 \text{ (m)}$$
$$H_2 = 10 \text{ (cm)} = 10^{-1} \text{ (m)}$$

$$\left(\frac{j_{2\rightarrow 1}}{j_{1\rightarrow 2}}\right)\left(\frac{H_1}{H_2}\right) = \left(\frac{10^{-4} f_{p2} + 10^{-1} f_{d2}}{10^0 f_{p1} + 10^{-1} f_{d1}}\right)(10^2)$$

$$= 10^{-2}(f_p \rightarrow 1) \rightarrow \text{Sediment}$$

$$= 10^{+2}(f_p \rightarrow 0) \rightarrow \text{Water column}$$

Source: From Di Toro DM, Paquin PR. 1984. Environ Toxicol Chem 3:335-353. With permission.

3.5.1 APPLICATION TO CLASSIFICATION

The aim of a classification system is to divide substances into classes according to their intrinsic hazards (for example, slightly, moderately, or highly toxic) for a given regulatory objective. Using an UWM approach, substances would be assigned into classes on the basis of their critical load. Current classification systems are most often based only on the aquatic compartment. Thus, it follows that the UWM would be first used for aquatic hazard assessment. When appropriate ecotoxicological data are available for all compartments, it would be possible to classify substances according to the most sensitive endpoint that triggers the critical load. For the purpose of classification on the basis of aquatic effects only, the water and sediment compartments of the model would be used. The classification of the substance would be carried out with a steady-state version of the model. The characteristics of the model system to be used must be standardized for all substances. Background levels of naturally occurring substances could be included in the modeling approach allowing for consideration within the classification system. Definition of the class boundaries might be assigned based on the magnitude of the critical load required to produce a critical effect, thus yielding classes of substances considered to be slightly, moderately, or highly toxic. This approach would be equally valid for organic substances.

3.5.2 APPLICATION TO RANKING

An additional aim of the UWM would be to produce a relative ranking of substances based on effects in the most sensitive environmental compartment. This model would result in a ranking of each substance on the basis of the critical load that would be required to trigger these effects. As for classification, the dimensions of the model are of lesser importance, but the system must be standardized to rank the substances against each other. An analogous approach has been developed for determining characterization factors for life cycle impact assessment (Huijbregts et al. 2001). Data gaps for ecological effects in a certain compartment would need to be addressed to consider all of the possible endpoints for all substances (modeled data could be considered for those compartments where experimental data are lacking). As for the use of the model in classification, the calculations should be performed by the application of a steady-state model. The contribution of background levels of a substance toward the total compartment levels may need to be considered to evaluate the additional loading (the critical load) required to trigger adverse effects. The ranking of chemicals would be based upon the predicted critical load required to trigger adverse effects.

3.5.3 APPLICATION TO SCREENING ASSESSMENT

The aim of a screening assessment is to identify which of the compartments may be the most sensitive to a given distribution of mass input for steady-state conditions. The UWM could be used in a time-dependent manner (nonsteady-state version of the model) to identify which compartment would be impacted first. It is anticipated that significant further development would be required for the proposed UWM to be adequately applied for risk assessment purposes, although future application for this purpose is a possibility.

3.5.4 DISTRIBUTION OF THE MASS INPUT INTO COMPARTMENTS FOR DIFFERENT ASSESSMENT LEVELS

The distribution of the mass input to the compartments of the multicompartment UWM would, to some extent, depend on the hazard assessment level (classification, priority ranking, and screening assessment) and its purpose. For assessment of the critical load for an overall hazard classification or ranking, separate simulations should be conducted with emissions to single compartments of air, water, and surface soil, that is, 3 separate modes of entry. If only the aquatic hazard is to be evaluated, only the single mode of entry to water needs to be addressed. Likewise, if only the terrestrial hazard is of concern, the mode of entry can be entirely to soil. A situation involving simultaneous inputs to more than 1 compartment can be considered, that is, multiple modes of entry. Assuming the fate equations are linear in concentration, the principle of superposition or linear additivity applies, and results for multiple modes of entry can be deduced by scaling and adding single mode of entry results. For metals, the use of superposition or linear additivity requires further evaluation, because metal partitioning behaves nonlinearly with respect to concentration. The distribution of the mass input to the different compartments should be designed on the basis of the objective of the screening.

3.5.5 PRECAUTIONARY APPROACH

The UWM approach presented in the previous section is most applicable when the effect levels, the modes of toxic action, and physicochemical properties of a substance are well known. Because of extensive data requirements for the substances to be assessed and limitations of the modeling framework, the proposed approach may not be initially applicable for all substances. In these cases, there may be a need to use other methods or information that could result in a chemical being assessed in a precautionary manner until the appropriate substance-specific data are obtained.

3.6 ILLUSTRATIONS OF HAZARD ASSESSMENTS

In the following sections, the general structure of the entire generic environment of the model (the Unit World) is further outlined, and some hypothetical results for 4 organic compounds and 2 metals are presented for illustrative purposes. Additionally, the soil compartment is examined and, finally, the water-column sediment compartments are examined in more detail, these being the most critical compartments from the toxicity viewpoint, at least for performing aquatic hazard assessments. Note that for these hypothetical results, soil is divided into surface and deep portions (soil is treated as a single compartment in Figure 3.1). The 6 compartments discussed are: air, water, surface soil, deep soil, surface sediment, and deep sediment.

3.6.1 EXAMPLE 1: THE GENERIC ENVIRONMENT (UNIT WORLD)

The UWM mass balance equations can be written in conventional concentration/partition coefficient/rate constant format as illustrated in Table 3.1 through Table 3.4 or they can be written in the fugacity/aquivalence format as given in Appendix A. Both methods should give identical results if equally parameterized. The illustrative results given later in Figure 3.3 through Figure 3.11 were obtained using the fugacity version.

Figure 3.1 illustrates the compartments in a hypothetical catchment basin. Key transport and transformation processes are indicated by arrows. Not all processes, or indeed all compartments, are necessary for all substances. There may be emissions or discharges to any or all compartments, but in practice, for metals, there is likely to be emission to the atmosphere (for example, from combustion or smelting sources), to the upper soil horizon, and to the water column. The precipitation rate into this hypothetical catchment is on the order of 1 m/year, that is, a total of 10^8 m^3 annually to an area of 10^8 m^2. The water outflow rate is thus approximately 10^4 m^3/h. The lake volume is 10^8 m^3; thus, the residence time of the water is about 10^4 hours or 1 year.

In principle, there is a separate loading or source evaluation as discussed in Section 3.3 that results in a continuous or pulse input of substance in a defined form to relevant compartments at specified rates in units such as mol/h or, if desired, in derived units such as g/m^2 year.

For illustrative purposes, the UWM is first applied to 4 hypothetical organic chemicals to illustrate the role of the model in translating critical toxic concentrations to critical emission rates (Section 3.6.1.1). It is then applied, again illustratively and hypothetically, to 2 metals (Section 3.6.1.2). These examples show that, in principle,

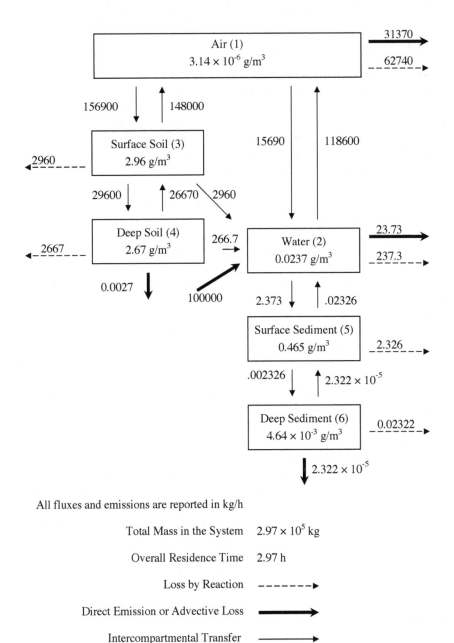

FIGURE 3.3 Mass balance diagram for organic chemical A in a 6-compartment unit watershed with emission of 100,000 kg/h into water.

it is possible to apply the same unit world mass balance approach to both organics and metals. A consistent evaluation approach is thus applicable to both.

3.6.1.1   Organic Compounds

Figure 3.3 gives a steady-state mass balance for a hypothetical organic compound A, which is fairly volatile. The emission rate is 100,000 kg/h, or 100 t/h to the water compartment, with a resulting concentration of 0.0237 g/m^3 or mg/l. The major loss process from water is evaporation. If a critical concentration in water is deemed to be 0.002 g/m^3 or 2 µg/l, that is, a factor of 11.86 lower than the value calculated above, this can be achieved by reducing the emission rate by this factor to 8428 kg/h or 8.43 t/h. In this case the equations are entirely linear, and a reduction in emission rate results in an exactly proportional reduction in all concentrations, masses, and fluxes. If nonlinearity exists, an iterative solution is required. Figure 3.4 shows this mass balance to achieve the critical concentration. The critical emission rate to water is, therefore, 8.43 t/h.

A similar analysis can be conducted for emissions to air and soil as illustrated in Figure 3.5 and Figure 3.6, yielding the same critical concentrations in water. Clearly, the critical emission rates to air and soil are significantly higher, namely 52.9 and 49.3 t/h, respectively. This illustrates the importance of mode of entry (i.e., proportions to each compartment, the partitioning and reactive properties of the substance, and its toxicity, which determine C_w — critical water concentration) in addition to the nature of the compartments and the flow parameters (which will presumably be the same for all substances evaluated). There is a marked difference in the efficiency with which the substance reaches the water depending on how it enters the Unit World. Efficiency is determined by the set of intermedia transport parameters and the rates of degrading reactions, all of which are chemical specific.

Figure 3.7 illustrates the mass balance for organic chemical B, a less volatile organic substance with the same critical water concentration. The critical emission rate to water is now considerably lower (0.12 t/h) because evaporation is retarded, and there is a build up of chemical in the water compartment.

Figure 3.8 illustrates the critical mass balance for organic chemical C, which has identical properties to B, but is less toxic, having a critical concentration in water of 0.02 g/m^3, that is, a factor of 10 higher than the previous substances. The critical emission rate to water is, as expected, a factor of 10 higher.

In general, higher critical concentrations result in higher critical emission rates, but this relationship is influenced by the transport efficiencies in the system. For example, in Figure 3.9 for organic chemical D, the critical concentration in water is 0.01 g/m^3, that is, it is more toxic than substance C, but because it is more volatile and reactive, a higher critical emission rate is tolerable. It is thus more toxic when viewed in terms of concentration in water, but it is less toxic when viewed in terms of critical emission rate because it is more rapidly lost from water.

In summary, if a UWM is fully defined, it is possible to calculate the critical emission rates to air, water, and soil that will yield critical concentrations in any of the receiving media. For example, critical concentrations could be defined in air, surface soil, water, or surface sediment. If desired, emissions could be to more than 1 com-

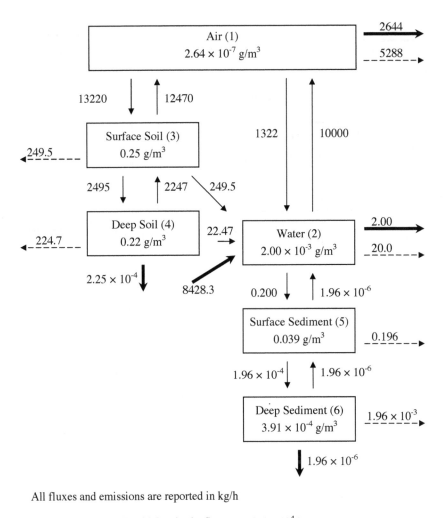

All fluxes and emissions are reported in kg/h

Total Mass in the System 2.5×10^4 kg

Overall Residence Time   2.966 h

Loss by Reaction $----->$

Direct Emission or Advective Loss $\longrightarrow$

Intercompartmental Transfer $\longrightarrow$

FIGURE 3.4 Mass balance diagram for organic chemical A in a 6-compartment unit watershed with emission reduced to achieve the critical concentration in water of 2 µg/l.

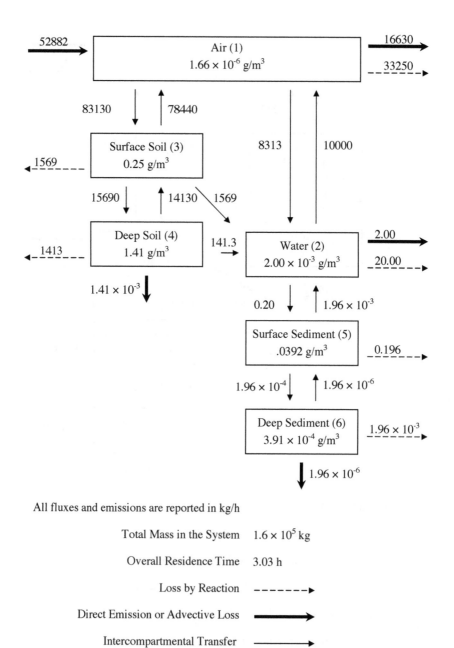

FIGURE 3.5 Mass balance diagram for organic chemical A in a 6-compartment unit watershed with emissions to air to achieve the critical concentration in water of 2 µg/l.

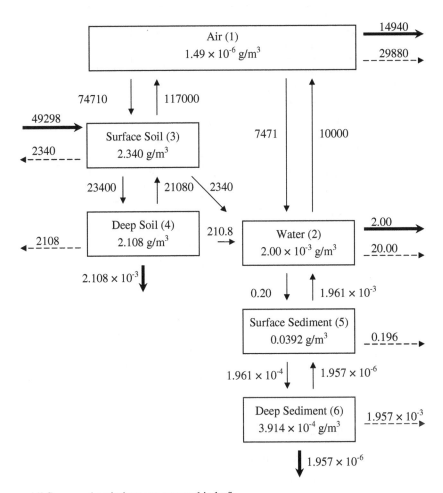

All fluxes and emissions are reported in kg/h

FIGURE 3.6 Mass balance diagram for organic chemical A in a 6-compartment unit watershed with emissions to soil to achieve the critical concentration in water of 2 μg/l.

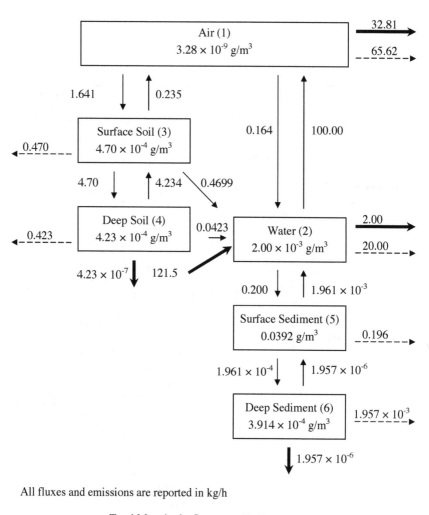

All fluxes and emissions are reported in kg/h

Total Mass in the System   92.50 kg

Overall Residence Time   0.761 h

Loss by Reaction $- - - - - - \blacktriangleright$

Direct Emission or Advective Loss $\longrightarrow$

Intercompartmental Transfer $\longrightarrow$

FIGURE 3.7 Mass balance diagram for organic chemical B in a 6-compartment unit watershed with emissions reduced to achieve the critical concentration in water of 2 μg/l.

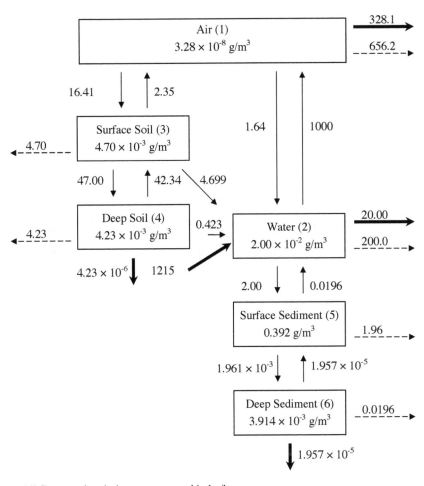

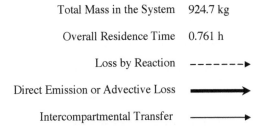

All fluxes and emissions are reported in kg/h

Total Mass in the System   924.7 kg

Overall Residence Time   0.761 h

Loss by Reaction $- - - - - \blacktriangleright$

Direct Emission or Advective Loss $\longrightarrow$

Intercompartmental Transfer $\longrightarrow$

FIGURE 3.8 Mass balance diagram for organic chemical C in a 6-compartment unit water-shed with emissions reduced to achieve the critical concentration in water of 20 µg/l.

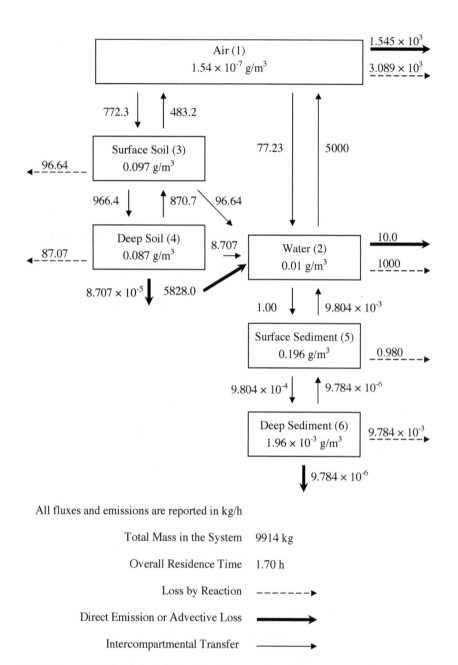

FIGURE 3.9 Mass balance diagram for organic chemical D in a 6-compartment unit watershed with emissions reduced to achieve a critical concentration in water of 10 μg/l.

partment, but interpretation is then more difficult. The concentrations can be total or dissolved in the case of water. Bioavailability, as influenced by sorption, is automatically taken into account if the critical concentrations are expressed as total quantities.

3.6.1.2 2 Metals

An analogous process is possible for metals, although nonlinearities are expected as a result of more complex sorption and dissolution processes than for organics. It may also be necessary to treat more than one ionic species. Degrading reactions can be set to zero. Sorption can be a controlling process for metals that sorb strongly to substrates such as sediments and soils. Strongly sorbing metals will be less efficiently transported in the water column or soil solution, depending on sorption to competing ligands such as humic and fulvic acids and will reside primarily in the sediment/soil compartment.

Figure 3.10 and Figure 3.11 give illustrative mass balance diagrams for 2 metals, M1 and M2, both of which are nonvolatile and nondegradable and for which emission is to soil. Metal M1 is strongly sorbed to soil and sediment. The principal route of loss is advection in water. Metal M2 is sorbed less, thus it leaches more rapidly, building up lower concentrations in soils and sediments. The principal route of loss is again advection from water. The degradation reactions used for the organic substances are now not relevant, although an irreversible aging removal process is included in soil, which acts against inputs from weathering.

The other significant difference between organics and metals is that the speciation of the metal in the water column is defined. For organics, only partitioning between the dissolved and particulate phase is needed, unless of course, the organic substances ionizes or speciates. The expressions describing speciation and sorption are considerably more demanding in the case of metals. Sorption is likely to be nonlinear in metals whereas it is linear for organic chemicals.

Another useful quantity that can be gleaned from these mass balances is the overall residence time or persistence of the substance. This is the ratio of the total mass of substance in the system to the rate of emission (or total loss) and is presented on each figure. This conveys information on how long it will take the system to reach steady state. Although the compartments approach steady state at different rates, the system is about 95% of the way to steady state after three residence times. If necessary, a dynamic model can be assembled and solved to characterize the approach to steady state more rigorously. This is likely to be more important for metals than for organics because the residence time of organics is usually controlled by degradation rates rather than burial or advective removal processes. However, constructing a dynamic model requires additional data, assumptions, and model complexity.

3.6.2 EXAMPLE 2: A SIMPLE APPROACH FOR SOILS

3.6.2.1 Defining a Unit World Soil

The Unit World approach for hazard assessment means that we want to define a soil system with attributes representative of the real world that allows us to consider the influences of the processes governing some or all of the fluxes defined in Section

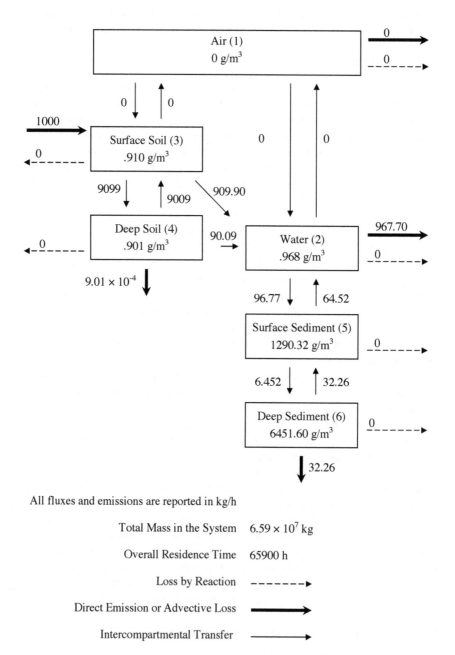

FIGURE 3.10 Mass balance diagram for metal M1 in a 6-compartment unit watershed with emission of 100,000 kg/h to surface soil.

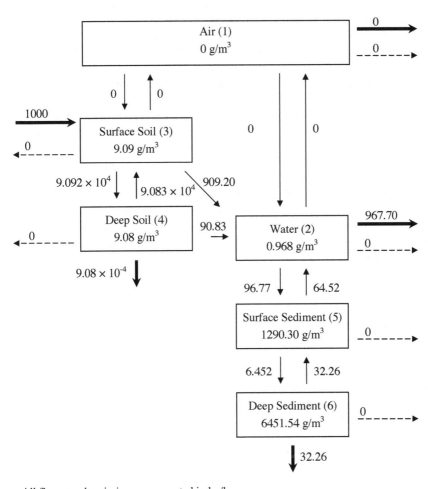

All fluxes and emissions are reported in kg/h

Total Mass in the System 6.68×10^7 kg

Overall Residence Time   66800 h

Loss by Reaction $------\blacktriangleright$

Direct Emission or Advective Loss $\blacksquare\blacksquare\blacksquare\blacktriangleright$

Intercompartmental Transfer $------\blacktriangleright$

FIGURE 3.11 Mass balance diagram for metal M2 in a 6-compartment unit watershed with emission of 100,000 kg/h to surface soil.

3.3.3. The following discussion lays out an approach that could be used to build a UWM for soils and gives a few working examples of simplified model calculations.

The model could be initially developed without accounting for the dependence of toxicity on chemistry, that is, the free ion approach, because we are dealing with a single set of conditions. This would allow for the use of extensive data sets, based on OECD standard soil that express toxicity as total reactive metal added to the soil. It is recognized that this is a conservative approach to estimating hazard and is intended for use in ranking or screening-level assessments. Therefore, we propose initially to use the OECD standard soil as the unit world soil, with additional definition of the composition of the soil water. The OECD standard soil comprises 70% sand, 20% kaolin, 10% peat, and has a pH of 6.0.

A key question is, which fluxes should be considered in the unit world soil system, given that simplicity is desirable and data availability may be restrictive? We suggest that the formation of metal precipitates may initially not be included during model development and harvesting flux be neglected by specifying the Unit World to represent natural or seminatural ecosystems. For most metals, removal in wind-blown dust would be nonspecific, because a very high fraction of total metal will be sorbed to the soil. Moreover, if the Unit World is assumed to have a confined atmosphere, dust will be redeposited, and the net flux will be 0. The same argument could be applied to volatilization. With these fluxes neglected, we are left with metal inputs from weathering, and losses by aging effects and drainage.

If weathering inputs are to be taken into account, the source minerals must be defined. If the minerals are taken to be those in the OECD standard soil, then it can be argued that the toxicity tests have already taken account of the background (weathered) metal, because this metal will have been present in the control tests. In other words, the toxicity endpoints refer to additions to the background, rather than the actual total reactive metal in the soil. Therefore, we can neglect the weathering flux in the Unit World.

Finally, we come to the aging term. Aging has been observed when metal is added to soil in experiments of short duration compared to geochemical processes, and which are therefore unlikely to represent steady-state conditions. The nature of the aging effect is not fully understood, but it might involve the migration of metal into mineral (especially oxide) matrices, followed by "trapping." During progression to a steady-state condition, this sink could become fully occupied and then its influence would cease, unless the trapping mineral could be generated by weathering at a significant rate, which seems unlikely. Therefore, for the conservative steady-state UWM, it seems justified to neglect the aging effect, at least until its nature and capacity can be defined for different metals.

By these arguments we can reduce the system to the very simple one in which metal output is because of only the drainage flux (Equation 3.5). After omitting the fluxes discussed above, combination of Equation 3.1, Equation 3.3, and Equation 3.5 gives:

$$\text{Critical Load} \quad = \quad F_{in,tox} \quad = \quad \frac{Q_{M,tox}}{K_D} w_F \qquad (3.6)$$

or, because w_F is a constant, we can define

$$\text{Haz}_M \;\; = \;\; \frac{w_F}{F_{in}} \;\; = \;\; \frac{K_D}{Q_{M,tox}} \tag{3.7}$$

where Haz_M is a measure of a metal's hazard; the larger its value, the greater the hazard. We thus come to the conclusion that the most hazardous metals have a high toxicity (low $Q_{M,tox}$) and a strong tendency to associate with soil solids. Note that Haz_M is simply the reciprocal of the steady-state aqueous metal concentration that corresponds to the toxic end point; the lower this concentration is, the more hazardous the metal.

3.6.2.2 Scoping Calculations

To get some idea of how this method would work in practice, we now perform some calculations for 4 cationic metals (Cu, Zn, Cd, and Pb) with differing toxicity and chemical characteristics. Table 3.5 shows toxic end points in the same test with the OECD standard soil. For simplicity, we assume that the only metal binding phase in the solid phase is the organic matter, that the aqueous phase has DOM but no SPM. The soil version of WHAM/Model VI (Tipping 1994, 1998) is then used to estimate the distribution of metal between the solid and aqueous phases, taking account of sorption and solution complexation, thereby yielding K_D (Equation 3.3). Values of Haz_M for different DOM concentrations are then derived; DOM is the principal aqueous phase complexant under the conditions of the unit world soil for the 4 metals under consideration.

The results in Table 3.5 indicate that whereas the metal toxicities increase in the order Pb ~ Cu ~ Zn < Cd, Haz_M increases in the order Zn < Cd < Cu < Pb. Thus, toxicity alone does not provide the same measure of hazard as the critical load approach, because of the differing chemistries of the 4 metals. The relative values

---

TABLE 3.5
Toxicity Data and Calculated K_D Values for the OECD Standard Soil

Metal	$Q_{M,tox}$ µmol g^{-1}	[DOM] = 1 mg l^{-1}		[DOM] = 10 mg l^{-1}		[DOM] = 100 mg l^{-1}	
		K_D l g^{-1}	Haz_M l µmol^{-1}	K_D l g^{-1}	H_M l µmol^{-1}	K_D l g^{-1}	H_M l µmol^{-1}
Cu	4.0	2300	580	1100	290	190	47
Zn	3.9	35	9.1	35	9.0	32	8.2
Cd	0.20	10	49	10	50	9	46
Pb	4.3	8400	1900	3100	730	430	100

Note: The toxicity data refer to *Folsomia*. From Sandifer RD, Hopkin SP. 1996. Chemosphere 33:2475-2486. With permission. The K_D values were calculated with WHAM/Model VI. H_M is given by Equation 3.7 (Section 3.6.2.1).

of Haz_M depend upon the value chosen for [DOM], because this affects the more strongly complexing metals (Cu and Pb) more than the weakly complexing ones (Zn and Cd).

By assuming the unit world to be in a steady state, no account is taken of the time to reach the toxic condition. For metals, such times can be extremely long. In the examples considered, for a 15 cm depth of initially metal-free soil, with 1 m a^{-1} runoff and [DOM] = 10 mg l^{-1}, it would take several thousand years for the soil Cd pool to approach steady state, and about 1 million years for the Pb pool to do so.

3.6.2.3 Application

The Haz_M approach could be applied immediately to all metals for which toxicity data are available for the OECD standard soil, and for which WHAM parameters are available, if organic matter is assumed to be the only sorbent and the principal aqueous phase complexant. To assess anionic metals, the modeling of K_D would require additional models to deal with interactions at oxide-type surfaces. An alternative to modeling the K_D values would be to determine K_D experimentally by equilibrating the soil, loaded with metal corresponding to $Q_{M,toc}$, with a standard soil water, and then determining total aqueous metal. In the examples shown here, data from only a single toxicity test were used; it would be preferable to combine data from a wider range of tests to give a better indication of hazard to the ecosystem, instead of to a single organism. The proposed approach is thus relatively straightforward, makes use of available data, and has the potential to be relatively easily implemented in cases where data are lacking.

3.6.3 EXAMPLE 3: THE WATER COLUMN/SEDIMENT MODEL

This final example highlights the differences that occur for different metals. The physical setting is a simple 2-layer water column/sediment model (Figure 3.12). No loss terms are included in the model, it is simply a lake (with no inlet or outlet) with a sediment. The water column and sediment interact via particles settling (w_1), resuspension (w_{12}), and pore water, overlying water diffusive exchange (K_{L12}). The parameters and equations used in the simulation are listed in Table 3.4. The only metal specific parameters are the partition coefficients in the water column and the sediment. Partitioning is assumed to be to particulate organic carbon (OC) only. Because the water-column particles are assumed to be 20% OC and the sediment is 2% OC, a usual case, the water-column partition coefficient is ten times that of the sediment. The model is simple enough that an analytical solution is available (Table 3.4).

The initial metal concentration in the water column is assumed to be $C_{T1}(0)$ = 1 μM. The total metal water-column concentration $C_{T1}(t)$ decreases exponentially until a steady-state concentration, $C_{T1}(0)(s_2/s_1 + s_2)$ is reached, which depends on the ratio of the magnitudes of the fluxes from and to the sediment, ($j_{2\rightarrow1}/j_{1\rightarrow2}$), and the ratio of the depths of the water column and sediment, (H_1/H_2). The resulting concentration ratios $C_{T1}(\infty)/C_{T1}(0) = (s_2/s_1 + s_2)$ are plotted vs. the sediment partition coefficient in Figure 3.13. The residence time in the water column is

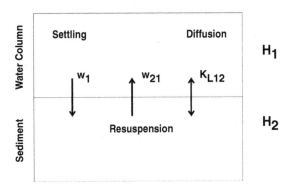

FIGURE 3.12 Water column/sediment model.

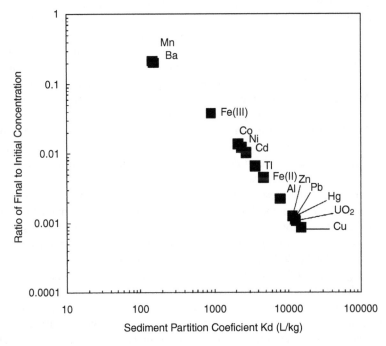

FIGURE 3.13 Ratio of final to initial metal concentration vs. sediment partition coefficient.

almost exactly inversely related to the sediment partition coefficient. The reason for this can be seen in the last equation of Table 3.4. The steady-state concentrations are functions of the particulate and dissolved fractions: f_{p1}, f_{d1}, f_{p2}, and f_{d2}, which are, in turn, functions of the partition coefficients. It is clear from the magnitude of the coefficients that the dissolved fraction in the pore water f_{d2} controls the steady-state concentration. This is because of the importance of the diffusive exchange rate. Because the partition coefficients span three orders of magnitude, so also does f_{d2}, and hence the steady-state solution. This example illustrates the importance of the partition coefficient in determining the fate of metals in the aquatic environment.

3.7 STEPS REQUIRED TO PROCEED FROM A PROTOTYPE TO A WORKABLE MODEL

3.7.1 IMPLEMENTATION

Although the evaluative approach described above is based on well-established principles governing the environmental behavior of metals, the proposed approach should still be considered at present as a conceptual framework embodying components that are at different stages of development and evaluation. For example, geochemical equilibrium models, such as MINTEQ2 (U.S. EPA 2000), which is used to compute the distribution in natural waters of dissolved metal species including their complexes, have a long history of development and evaluation and are widely used. On the other hand, the adsorption of metals on soils and sediments is not as well characterized, because metals can adsorb to a variety of incompletely defined sorbent solids, such as clay minerals, organic matter, and oxyhydroxides of Fe and Mn. At the present time, greater progress has been made in developing predictive approaches for addressing the sorption of metals to oxic sediments (Tipping 1994) than to soils. Methods for assessing the formation of insoluble metal sulfides in sediments have progressed rapidly in recent years (Di Toro et al. 1996). Progress has also been made in developing models that estimate the fate of some metals in surface and groundwater systems (Diamond et al. 2000; Di Toro 2001; Mayer et al. 2002), but these approaches are limited to describing the transport of specific metals under site-specific conditions. Progress has also been made in developing a generic model for metals in aquatic systems but is limited by full descriptions of processes for a wider range of metals and metalloids and system characteristics (Bhavsar et al. 2004a, 2004b). It will, therefore, be necessary to undertake a series of well-integrated activities in order to move forward from the conceptual stage to a fully implemented and accepted evaluative method that is capable of tracing the significant fate and transport processes of a wide range of metals and predicting both the concentration and speciation at the exposure point with a sufficient degree of accuracy to reflect the objectives of the assessment, for example, classification, ranking, or screening-level assessment. Also, the integration of individual models will require that close attention be paid to harmonizing assumptions to ensure consistent treatment of processes and phenomena. The first step in this process should entail the construction of a prototype model. Initially, this model can be built by integrating existing modeling frameworks that individually have been designed to address one aspect or another of the overall model; however, refinements will be necessary as previously noted.

The next step will entail applying the model to a number of existing data sets for the purpose of calibration and evaluation. Several data sets could be used for this purpose (e.g., Di Toro 2001 and references therein; Mayer et al. 2002; Bhavsar 2005). These data sets have been obtained in aquatic mesocosms and in environmental systems and include several metals. In verifying the accuracy of the UWM's performance over a range of metals and evaluative environments, there are several considerations that will need to be addressed, as noted in Section 3.7.2.

3.7.2 UWM EVALUATION AND VALIDATION

Inevitably, when mass balance models are used for regulatory purposes, the question is asked, "has the model been validated against real environmental data?" The issue of testing environmental models has been discussed in detailed by Oreskes et al. (1994). Their general conclusion was that models that represent complex environmental phenomena cannot be validated, if one defines validation as establishing the "truth." This situation arises because the models represent open systems with numerous measured input parameter values that provide output for that known condition. It is not possible to devise an independent and rigorous test of the model under a variety of circumstances in which all model parameter values and test data can be known. Also, because of system complexity, circumstances can always be identified in which the model will fail to give reasonable results in some or all compartments. Another consideration is that the degree of correspondence between measured and modeled observations of a simple or complex multimedia environment, and expectations are sensitive to the use of the model. Because of all these factors, Oreskes et al. (1994) recommended the more general term "evaluation" be used when dealing with environmental models, and cautioned that we can only have limited confidence in model performance as a result of comparison exercises.

In light of the "open ended" nature of model evaluation, 3 less ambitious goals can be pursued. First, the model can be verified, which connotes that the mathematical expressions and their translation to a computer code truly represent the intention of the modeler and that all details of the calculation are "correct." In this case, transparency of the computer code is essential. Second, the model can be tested for consistency with results obtained from controlled systems, such as mesocosms in which the fate of a known quantity of metal can be monitored under closely defined conditions. The limitation of this approach is that conditions in mesocosms differ from those in the environment and, as such, this is an approximate but not complete evaluation of model performance under true environmental conditions. Third, individual equations and parameters used to describe a process may be shown to be sound when tested under laboratory or field conditions. This lends credibility to the formulation but, again, this evaluation does not provide a test of the whole model because of uncertainties associated with scaling up from single-process measurements taken at specific locations and points in time, to estimate this process at a whole system level over a longer time frame.

The best that can be achieved is the assembly of a set of verified, robust, well-tested and transparent equations and parameters that have acquired credibility from a range of tests and test circumstances. This assemblage, constituting the entire model, then requires further examination at the whole system level. It represents the best expression of the state-of-the-art of understanding and quantifying system behavior.

It must be emphasized that the proposed model is evaluative; it is not a simulation model, although it contains many of the same equations and parameters. Rather, it represents an attempt to describe the approximate and average fate of metals under defined, hypothetical, and representative conditions that may not exist at any one place or time. In this sense, its intended use is similar to that of the European Union System for the Evaluation of Substances (EUSES) (EC 1996).

A problem remains of establishing credibility through model evaluation and ensuring that current scientific understanding of the critical processes is included (note that models do not necessarily strive for completeness by including all known processes, but rather focus on the dominant processes that determine metal fate). This is best done by impartial peer review, application of the UWM to existing data sets or sponsorship of studies to collect data sets that can be used to evaluate the UWM and, finally, comparisons among independent models that purport to describe the same phenomena. Undoubtedly, the UWM and associated developments will evolve as the science progresses and, thus, the UWM must be regarded as a dynamic instrument by which the current scientific understanding is translated into a regulatory tool. For example, for organic substances, an air or atmosphere compartment would be required. The UWM cannot be validated, but it can be evaluated and it should ultimately prove to be a useful tool for regulatory purposes. Finally, activities will need to be undertaken to reach consensus on a harmonized approach to application of the method for the purposes of classification, ranking, and screening-level assessment. Issues that will need to be addressed include agreement on evaluative environmental settings and their characteristics, appropriate values for the various metal-specific properties required in the model, and applicable decision frameworks. If necessary, further data generation and research to refine the method will need to be undertaken, including:

- Discussions among the modelers to finalize the model structure, dealing with details such as the source term, how to adequately capture background metal levels, choice of partitioning functions for soils (free metal ion/total soil metal), compartment volumes, and default values.
- Consideration of data needs, and decisions about modifications in view of data absence, or determining missing data. This will be especially difficult for metals that are little-studied, such as thallium.
- Determination of processes and parameter values, which are both highly uncertain and critical to model results. An example is the procedure to quantify the loading term for metals in commerce and the behavior of redox-sensitive metals such as arsenic and selenium. A number of virtual watersheds need to be determined that are representative of a realistic range of conditions and the UWM used for trial runs to see how different assumptions affect the hazard ranking.

ACKNOWLEDGMENTS

We acknowledge contributions from Bill Wood (U.S. EPA, Washington, D.C.), Johanna Peltola-Thies (Federal Environmental Agency, Berlin), and John Westall (Oregon State University, OR) during and following the initial workgroup meetings in Pensacola, FL.

REFERENCES

Adams WJ, Conard B, Ethier G, Brix KV, Paquin PR, DiToro DM. 2000. The challenges of hazard identification and classification of insoluble metals and metal substances for the aquatic environment. Human Ecol Risk Assess 6:1019–1038.

Bhavsar SP. 2005. Development of a coupled metal transport-speciation model for surface aquatic systems and its extension to soil. Ph.D. Dissertation, Department of Chemical Engineering and Applied Chemistry, University of Toronto, ON, Canada.

Bhavsar SP, Diamond ML, Evans LJ, Gandhi N, Nilsen J, Antunes P. 2004a. Development of a coupled metal speciation-fate model for surface aquatic systems. Environ Toxicol Chem 23:1376–1385.

Bhavsar SP, Diamond ML, Gandhi N. 2004b. Dynamic coupled metal TRANSport-SPECiation (TRANSPEC) model: application to assess a zinc contaminated lake. Environ Toxicol Chem 23:2410–2420.

Bhavsar SP, Gandhi N, Diamond ML, Lock AS, Spiers G, Aforo de la Torre MC. 2006. Effects of estimates from WHAM and MINEQL+ on metal fate predicted by TRANSPEC — a coupled speciation-fate model. In preparation.

Blowes DW, Gillham RW, Ptacek CJ, Puls RW, Bennett TA, O'Hannesin SF, Hanton-Fong CJ, Bain JG. 1999. An in situ permeable reactive barrier for the treatment of hexavalent chromium and trichloroethylene in ground water: volume 1, design and installation. U.S. Environmental Protection Agency, EPA/600/R-99/095a, 111 p.

Blowes DW, Ptacek CJ, Jambor JL, Weisener CG. 2003. The geochemistry of acid mine drainage. In: Holland HD, Turekian KK, editors. Environmental geochemistry 9: treatise on geochemistry. Oxford, UK: Elsevier-Pergamon, p. 149–204. .

Boudreau BP. 1991. Modeling the sulfide-oxygen reaction and associated pH gradients in porewaters. Geochim Cosmochim Acta 55:145–159.

Bryan SE, Tipping E, Hamilton-Taylor J. 2002. Comparison of measured and modelled copper binding by natural organic matter in freshwaters. Comp Biochem Physiol C 133:37–49.

Chapra SC. 1997. Surface water-quality modeling. New York: McGraw-Hill.

Crommentuijn T, Polder M, Sijm D, de Bruijn J, van de Plassche E. 2000. Evaluation of the Dutch environmental risk limits for metals by application of the added risk approach. Environ Toxicol Chem 19:1692–1701.

De Vries W, Bakker DJ. 1998. Manual for calculating critical loads of heavy metals for terrestrial ecosystems. Guidelines for critical limits, calculation methods and input data. Report 166. Wageningen, The Netherlands: DLO Winand Staring Centre.

Diamond ML. 1995. Application of a mass balance model to assess in-place arsenic pollution. Environ Sci Technol 29:29–42.

Diamond M, Ganapathy M, Petersen S, Mach C. 2000. Mercury dynamics in the Lahontan Reservoir, NV: application of the QWASI fugacity/aquivalence multispecies model. Water Air Soil Pollut 117:133–156.

Di Toro DM. 2001. Sediment flux modelling. New York: John Wiley & Sons, Inc.

Di Toro DM, Paquin PR. 1984. Time variable model of the fate of DDE and lindane in a quarry. Environ Toxicol Chem 3:335–353.

Di Toro DM, Paquin PR. 2000 Persistence of metals. London, UK: International Council for Mining and Metals.

Di Toro DM, Mahony JD, Hansen DJ, Berry W. 1996. A model of the oxidation of iron and cadmium sulfide in sediments. Environ Toxicol Chem 15:2168–2186.

Doyle PJ, Gutzman DW, Sheppard MI, Sheppard SC, Bird GA, Hrebenyk D. 2003. An ecological risk assessment of air emissions of trace metals from copper and zinc production facilities. Human Ecol Risk Assess 9:607–636.

EC (European Commission). 1996. EUSES documentation — The European Union system for the evaluation of substances. National Institute of Public Health and the Environment (RIVM), The Netherlands. Available from the European Chemicals Bureau (EC/DGXI, Ispra).

EU (European Union). 1991. Council Directive amending for the seventh time Directive 67/548/EEC on the approximation of laws, regulations and administrative provision relating to the classification, packing and labeling of dangerous substances (92/32/EEC). Official Journal of the European Communities, June 5, 1991.

Existing Substances Branch. 2003. Guidance document for the categorization of organic and inorganic substances on Canada's domestic substances list: determining persistence, bioaccumulation potential and inherent toxicity to non-human organisms. Environment Canada. Hull, PQ, Canada.

Frau F. 2000. The formation-dissolution-precipitation cycle of melanterite at the abandoned pyrite mine of Genna Luas in Sardinia, Italy: environmental implications. Mineral Mag 64:995–1006.

Huijbregts MAJ, Guinee JB, Reijnders L. 2001. Priority assessment of toxic substances in a life cycle assessment. III: export of potential impact over time and space. Chemosphere 44:59–65.

Kleka G, Boethling B, Franklin J, Grady L, Graham D, Howard PH, Kannan K, Larson R, Lipnick R, Jansson B, Mackay D, and others, editors. 2000. American Chemical Society Symposium Series No. 773: persistent, bioaccumulative, and toxic chemicals, volume II. Washington, D.C.: Washington and Oxford University Press.

Lipnick RL, Janson B, Mackay D, Petreas M, editors, Persistent, bioaccumulative and toxic chemicals II, Washington, D.C.: ACS.

Lofts S, Tipping E. 1998. An assemblage model for cation binding by natural particulate matter. Geochim Cosmochim Acta 62:2609–2625.

Lofts S, Spurgeon DJ, Svendsen C, Tipping E. 2004. Deriving soil critical limits for Cu, Zn, Cd, and Pb: a method based on free ion concentrations. Environ Sci Technol 38:3623–3631.

Mackay D, Paterson S, Cheung B, Neely WB. 1985. Evaluating the environmental behavior of chemicals with a level III fugacity model. Chemosphere 14:335–374.

MacKay D, Shiu WY, Ma KC. 1992. Illustrated handbook of physical-chemical properties and environmental fate of organic chemicals. Vol. II Polynuclear aromatic hydrocarbons, polychlorinated dioxins and dibenzofurans. Chelsea, MI: Lewis Publishers.

Mackay D, Di Guardo A, Paterson S, Cowan CE. 1996. Evaluating the environmental fate of a variety of types of chemicals using the EQC model. Environ Toxicol Chem 15:1627–1637.

Mackay D, Webster E, Woodfine D, Cahill T, Doyle P, Couillard Y, Gutzman D. 2003. Towards consistent evaluation of the persistence of organic, inorganic and metallic substances. Human Ecol Risk Assess 9:1445–1474.

Mackay D, Hubbarde J, Webster E. 2003b. The role of QSARs and fate models in chemical hazard and risk assessment. Quant Struct Act Relat Comb Sci 22:106–112.

Mason AZ, Jenkins KD. 1995. Metal detoxification in aquatic organism. In: Tessier A, Turner DR, editors. Metal speciation and bioavailability in aquatic systems. Chichester, UK: John Wiley and Sons, p. 449–608.

Mayer KU, Frind EO, Blowes DW. 2002. Multicomponent reactive transport modeling in variably saturated porous media using a generalized formulation for kinetically controlled reactions. Water Resour Res 38:1174–1195.

McGeer JC, Brix KV, DeForest DK, Brigham SI, Skeaff JM, Adams WJ, Green A. 2003. Bioconcentration factor for the hazard identification of metals in the aquatic environment: a flawed criterion? Environ Toxicol Chem 22:1017–1037.

Nordstrom DK, Southam G. 1997. Geomicrobiology of sulfide mineral oxidation. In: Banfield JF, Nealson KH, editors. Geomicrobiology: interactions between microbes and minerals, vol. 35. Mineralogical Society of America, p. 361–385.

Nordstrom DK, Alpers CN. 1999. Negative pH, efflorescent mineralogy, and the challenge of environmental restoration at the Iron Mountain Superfund site, California. Proc Natl Acad Sci USA 96:3455-3462.

Nriagu JO. 1990. Global metal pollution. Environment 32:7–11, 28–33.

OECD (Organization for Economic Cooperation and Development). 2001a Harmonized integrated classification system for human health and environmental hazards of chemical substances and mixtures. OECD ENV/JM/HCL(2001)6, Paris, France. Available from: http://www.olis.oecd.org/olis/2001doc.nsf/LinkTo/env-jm-mono(2001)6.

OECD. 2001b. OECD series on testing and assessment. Guidance document on transformation/dissolution metals and metal compounds in aqueous media. Number 29. ENV/JM/MONO(2001)9. Paris, France. Available from: http://www.olis.oecd.org/olis/2001doc.nsf/LinkTo/env-jm-mono(2001)9.

Oreskes N, Shrader-Frechette K, Belitz K. 1994. Verification, validation, and confirmation of numerical models in the earth sciences. Science 263:641–646.

Pedersen TF. 1993. The early diagenesis of submerged sulphide-rich mine tailings in Anderson Lake, Manitoba. Can J Earth Sci 30:1099–1109.

Sandifer RD, Hopkin SP. 1996. Effects of pH on the toxicity of cadmium, copper, lead and zinc to *Folsomia candida* Willem, 1902 (*Collembola*) in a standard laboratory test system. Chemosphere 33:2475–2486.

Schnoor JL. 1996. Environmental modeling. New York: Wiley Interscience.

Skeaff JM, Delbeke K, Van Assche F, Conard B. 2000. A critical surface area concept for acute hazard classification of relatively insoluble metal-containing in aquatic environments. Environ Toxicol Chem 19:1681–1691.

Skeaff JM, Dubreuil AA, Brigham, SI. 2002. The concept of persistence as applied to metals for aquatic hazard identification. Environ Toxicol Chem 21:2581–2590.

Struijs J, Van de Meent D, Pijnenburg WJGM, Van den Hoop MAGT, Crommentuijn T. 1997. Added risk approach to derive maximum permissible concentrations for heavy metals: how to take into account the natural background levels? Ecotox Environ Saf 37:112–118.

Thomann RV, Mueller JA. 1987. Principles of surface water quality modeling and control. New York: Harper.

Tipping E. 1994. WHAM — a chemical equilibrium model and computer code for waters, sediments and soils incorporating a discrete site/electrostatic model of ion-binding by humic substances. Comp Geosci 20:973–1023.

Tipping E. 1998. Humic ion-binding model VI: an improved description of the interactions of protons and metal ions with humic substances. Aquat Geochem 4:3–48.

UN (United Nations). 2003. The globally harmonized system of classification and labelling of chemicals. ST/SG/AC.10/30. Available from: http://www.unece.org/trans/danger/publi/ghs/ghs_rev00/00files_e.html.

USEPA (U.S. Environmental Protection Agency). 2000. MINTEQA2 — metal speciation equilibrium model for surface and ground water. Washington, D.C.: Center for Exposure Assessment Modeling.

4 Bioaccumulation: Hazard Identification of Metals and Inorganic Metal Substances

Christian E. Schlekat, James C. McGeer,
Ronny Blust, Uwe Borgmann, Kevin V. Brix,
Nicolas Bury, Yves Couillard, Robert L. Dwyer,
Samuel N. Luoma, Steve Robertson,
Keith G. Sappington, Ilse Schoeters,
and Dick T.H.M. Sijm

4.1 INTRODUCTION

Bioaccumulation is the process whereby aquatic organisms accumulate substances in their tissues from water and diet. Bioaccumulation is of potential concern both because of the possibility of chronic toxicity to the organisms accumulating substances in their tissues and the possibility of toxicity to predators eating those organisms.

The objectives of this chapter are to review the regulatory tools that apply to bioaccumulation, to summarize the current knowledge on metal bioaccumulation processes, and to propose scientifically defensible approaches for fulfilling the regulatory intent of the use of bioaccumulation data. The chapter is divided into 6 sections. Section 4.2 reviews the rationale behind the regulatory concern over bioaccumulation and the use of various bioaccumulation indices by 3 regional regulatory agencies (United States, Canada, and Europe). Section 4.3 briefly introduces the mechanisms of metal bioaccumulation and the current understanding of the relationship between bioaccumulation and toxicity. Section 4.4 identifies the scientific rationale for considering that certain commonly used bioaccumulation indices do not fulfill the regulatory intent of bioaccumulation, and begins to identify how alternative approaches can be developed. Section 4.5 provides examples of how current scientific knowledge of bioaccumulation may be used to relate it to toxicity and identifies the limitations of these relationships. Section 4.6 discusses how bioaccumulation of different metals can be compared by incorporating bioaccumulation models into the UWM. Bioaccumulation models estimate tissue metal concentrations, and these concentrations can be compared to threshold dietary toxicity values. Section 4.7 provides the conclusions.

4.2 REGULATORY OBJECTIVES OF BIOACCUMULATION IN HAZARD ASSESSMENT

Brief examples of regulatory applications of bioaccumulation are provided for the European Union, the United States, and Canada in Section 4.2.1, Section 4.2.2, and Section 4.2.3, respectively.

The potential for a substance to bioaccumulate has been used as a surrogate for chronic effects in regulatory systems (OECD 2001). Traditionally, bioconcentration (i.e., uptake from water only) has been assessed using standard bioconcentration tests, where organisms are exposed to a substance in water and the resulting tissue concentrations are measured. The ratio of these values is the bioconcentration factor (BCF) (OECD 1996). Alternatively, bioaccumulation (that is, uptake from all media including water, food, and sediment) has been assessed by determining the ratio of chemical concentrations in organisms to that in water in natural ecosystems; this ratio is expressed as the bioaccumulation factor (BAF). Such data are not easily generated in the laboratory, and are, therefore, typically derived from field monitoring studies where colocated water and tissue concentrations are available. These bioaccumulation measures, along with the octanol–water partition coefficient (K_{ow}) for nonpolar organic compounds that are poorly metabolized, are highly valuable when little or no long-term toxicological data are available (OECD 2001). However, limitations to this approach exist for metals and are discussed below.

4.2.1 EUROPEAN UNION (EU)

Activities of the EU regarding hazardous chemicals include hazard assessments, risk assessments, and setting of environmental quality standards (for example, for water, groundwater, and sediment). In addition, the EU New Chemicals Policy (REACH: Registration, Evaluation, Authorization, and Restriction of CHemicals) will necessitate authorization for use of organic substances that are classified as PBT and vPvB (very persistent and very bioaccumulative). The low K_{ow} cut-offs for bioaccumulative and very bioaccumulative substances are 2000 l/kg and 5000 l/kg, respectively. Evaluation of metals for bioaccumulation potential in these frameworks also includes risk assessment and setting environmental quality standards, but is currently not performed in formal persistence, bioaccumulation, and toxicity (PBT)-assessments or hazard classification because of the recognition that, for metals, information other than BCFs should be used to assess bioaccumulation hazard (OECD 2001).

4.2.2 UNITED STATES

The U.S. Environmental Protection Agency (EPA) evaluates bioaccumulation information for classifying and prioritizing chemical hazard in several regulatory programs (e.g., the Toxics Release Inventory [TRI], the Hazardous Waste Minimization Prioritization Program [WMPT], and the New Chemicals Premanufacture

Notification Program). The general goal of these programs is to classify or rank large numbers of chemicals (hundreds to thousands) by selected attributes of interest (for example, persistence, bioaccumulation, and toxicity) for establishing priorities for future actions, such as setting release reporting requirements (e.g., TRI), or pollution prevention activities (e.g., WMPT). Classifying or ranking chemicals by their bioaccumulative properties is conducted by comparing aquatic-based BCF and BAF data to numeric benchmarks established by policy. For example, the TRI program uses a benchmark value of 1000 to classify a compound as bioaccumulative and a value of 5000 to classify a substance as highly bioaccumulative (EPA 1999a). As part of the WMPT, a bioaccumulation score of 1, 2, or 3 is assigned to chemical substances with BCF or BAF values of >250, 250 to 1000, and >1000. Because of complications associated with assessing metals' risk and hazards in a variety of contexts, the EPA is currently developing a comprehensive Metals Assessment Framework and Guidance for Characterizing and Ranking Metals (EPA 2002a). Because of this ongoing effort for improving metals' assessment procedures, the PBT scoring approach is not currently being applied to metals as part of the WMPT.

4.2.3 CANADA

Environment Canada has initiated a systematic categorization of the 23,000 substances on its Domestic Substances List (DSL). Categorization is not a process of hazard classification but rather a hazard-based priority-setting exercise. All the substances meeting prescribed criteria (according to the regulations) for persistence, or bioaccumulation, and inherent toxicity will be categorized and, subsequently, will be the object of a screening for ecological risk assessment. The DSL has to be categorized within a 7-year time frame that commenced on September 14, 1999 (CEPA 1999). Environment Canada has adapted the PBT framework for the categorization of metals and metal-containing inorganics. According to this modified scheme, all the metal-containing substances are considered by default as persistent and bioaccumulation is not used (it is considered as requiring further research). Consequently, inherent toxicity is the key discriminating factor (Borgmann et al. 2005).

4.3 SCIENTIFIC BASIS OF METAL BIOACCUMULATION: CURRENT STATE OF UNDERSTANDING

4.3.1 MECHANISMS OF METAL UPTAKE

Metal uptake in aquatic organisms occurs across the membranes that separate the organism from the external environment (Simkiss and Taylor 1995). In multicellular organisms, uptake is largely restricted to specialized organs such as the gills, in the case of waterborne uptake, and the digestive tract, in the case of dietary uptake. Most metal species that form in aquatic solutions are hydrophilic and do not permeate the membranes of these epithelia by passive diffusion. This means that the uptake of metals largely depends on the presence of transport systems that provide biological gateways for the metal to cross the membrane. This is in contrast to neutral organic substances, which are lipophilic and hydrophobic, and accumulate in biota via simple passive diffusion as predicted by Fick's Law (McKim 1994). Although metal uptake is usually via specific transport systems, there are exceptions, for example, some organometallic species such as tributyltin (TBT) compounds, or methylmercury, which behave like nonpolar organics and are taken up across the membrane by passive diffusion (Campbell 1995).

Most of the metal transport proteins present in biological membranes are involved in ion regulatory processes and the uptake of essential elements. Some of these transporters are highly selective for a single type of ion, whereas others are less selective and facilitate the uptake of different elements and species. For example, epithelial proteins involved in the transport of free iron, copper, and zinc ions may also carry nonessential elements such as cadmium or silver (Bury et al. 2003). Another example is the calcium ion channels present in the apical membranes of gill and other epithelia that can take up both Ca^{2+} and Cd^{2+} (Verbost et al. 1987) because of similarities in their charge and ionic radius.

Another important aspect of metal uptake and bioaccumulation is that uptake processes are complex and provide for dramatically different uptake (and elimination) processes along the spectrum of exposure concentrations. In the case of essential elements, for example, uptake across membranes can be via a number of different transport proteins, each with a unique affinity and capacity for the metal. To meet nutritional needs in times of deficiency, organisms activate physiologically-based feedback mechanisms that result in changes to the affinity/capacity of a transport protein or the relative number of particular proteins (e.g., low capacity–high affinity), available for uptake within a specific membrane system (Collins et al. 2005). Similarly, upon exposure to metal excess, in the short term, organisms may acclimate by decreasing metal uptake (McDonald and Wood 1993), although in the long term, the evolutionary pressure of high background metal concentrations may lead to adaptation (Klerks 2002). Consequently, metal uptake from the environment can be a function of the exposure concentration, the geochemical form, the biology of the species, physiological mechanisms, and interactions among these factors.

4.3.2 GILL VS. GUT ENVIRONMENTS

Metal uptake mainly occurs via the gills and the digestive system in aquatic organisms. Although the organization of these 2 systems is very different, they both include a variety of metal transporters. An important difference for metal uptake between these 2 systems is the nature of the gill and gut environment. The gill environment reflects the composition of the external solution to a certain extent although gradients in proton and other ion concentrations exist (Playle and Wood 1989). The gut environment differs more strongly from the external environment because of the active secretion of digestive fluids and enzymes in the lumen (Chen et al. 2002; Wilson et al. 2002). In addition, the functional organization of the digestive system shows important differences across species both within and among groups. In higher organisms such as fish, digestion is largely extracellular, but many invertebrates exhibit intracellular digestion involving the uptake of particulate matter across the apical membranes of the epithelial cells by endocytosis and further metabolic processing. The intestine is also the site of small organic molecule uptake. Metals may bind to these molecules and inadvertently enter tissues via these small organic molecule transporters (Vercauteren and Blust 1996; Glover et al. 2003). These various processes have very important consequences for the chemical speciation and biological availability of metals present in the ingested material (see Section 4.3.3).

4.3.3 CHEMICAL SPECIATION AND BIOLOGICAL AVAILABILITY

Metals occur in the aquatic environment under a variety of forms and species. It is well established that the speciation of a metal has an important impact on its uptake in biological systems (Campbell 1995). For uptake via the water phase it appears that, in most cases, the free metal ion is more readily available and taken up, although there are a number of significant exceptions. However, other factors such as dissolved organic carbon, water hardness, and hydrogen ion activity also have to be taken into account. These factors not only have a strong effect on the chemical speciation of metals, but they may also interact with metal transport proteins in a competitive (e.g., calcium ion) or noncompetitive manner (e.g., hydrogen ion) (Chowdhury and Blust 2001). The effects of these factors on metal uptake have been studied for a variety of species and conditions, and it has been shown that a relative simple metal uptake model, for example, a Michaelis–Menten model, can accommodate most of these effects.

Metal uptake from the diet is highly complex, as it occurs from a lumen environment that can be very different from that of the waterborne exposure solutions. As discussed in Section 4.3.2, the functional organization of the digestive system shows important differences among organisms both within and among groups and, therefore, the biological availability of metals from ingested food or sediment will vary with the organism considered, resulting in differences in assimilation efficiency. A detailed review of dietary metal uptake, organismal differences, and digestive processes has recently been published (Campbell et al. 2005). The diet is a major source of nutritive metals for most organisms. Consequently, organisms require well-regulated uptake processes to ensure a fine balance between deficiency and toxicity, particularly for nutritionally essential elements. The digestive processes (i.e., enzymes, acidity, redox, and retention time) are designed to liberate metal so that it is repackaged to the extent that it is recognized by the transport epithelium. Consequently, regulation of uptake primarily occurs at this epithelial membrane by the expression pattern of the transport proteins, complexation by mucus, or storage in the intestinal tissue.

A complicating factor in predicting the potential for metals to bioaccumulate from the diet is that they occur in a variety of forms and concentrations (e.g., algal cells, suspended and sediment particles, and prey items). For example, metal in prey species may exist in different forms depending upon the detoxification strategy of the prey organism (Rainbow 2002). Prey organisms that use metal granular formation as a detoxification mechanism (e.g., mollusks and some polychaetes) can reduce trophic transfer, because most of the metal appears inaccessible to the digestive process (Nott and Nicolaidou 1990, 1993; Wallace et al. 1998). However, predatory snails have been shown to assimilate relatively high proportions (40 to 80%) of metals associated with metal-rich granules formed by oysters that are preyed upon by the snails (Cheung and Wang 2005). Those organism that use cysteine-rich compounds for detoxification may increase trophic transfer due to the ease with which metals become liberated in the digestive process. Within this context, it is also important to consider the effect of the digestive process on the availability of metal species such as the metal sulfides that are present in anaerobic sediment layers. Although metals associated with sulfides are generally not available to infaunal organisms via pore water exposure, they can be assimilated with varying efficiencies via sediment ingestion (Lee et al. 2000). In marine copepods, bivalves, and larval fish, assimilation efficiencies of essential and nonessential metals have been shown to be directly related to the algal cytoplasm concentration of that metal (Wang and Fisher 1996; Reinfelder et al. 1998). In spite of this, links between subcellular metal fractions in a food item and metal assimilation should be considered with caution as other studies have shown that cytoplasmic metals either overestimate (Schlekat et al. 2000) or underestimate (Schlekat et al. 2002) assimilation efficiency.

4.3.4 BIOACCUMULATION AND TOXICITY

Once metals have translocated across the exchange epithelia, they may be compartmentalized within different organ compartments. Distribution among organs is variable depending on the site of exposure (gill vs. gut), the metal, and the mechanisms by which the metal integrates with the physiology of the animal. The bioreactive pool includes metals that can be incorporated in metabolically active molecules and participate in different types of physiological processes. Several families of evolutionary conserved proteins are involved in delivering essential metals to the appropriate cellular compartment for insertion into the correct cellular biological active unit (e.g., enzymes, DNA transcription factors — Huffmann and O'Halloran [2001]). Interestingly, the identification of these pathways has questioned the notion of a free metal ion pool in cells under normal conditions (Finney and O'Halloran 2003). However, toxicity is expected to occur when the concentration of the bioreactive pool exceeds a certain threshold level so that essential functions are impaired (e.g., inhibition of enzymes or transporters by binding of metals in the catalytic centre of the molecule). When the rate of metal uptake exceeds the rate of either elimination or detoxification, metal will accumulate in the bioreactive pool, and toxicity can occur when a threshold level is exceeded. This spillover theory for toxicity and some of the variations in storage, excretion, and internal regulation of metals that have been identified in marine organisms are shown with a series of schematic diagrams (adapted from Rainbow 2002) and presented in Figure 4.1. The potential for toxicity to be expressed is dependent on the relative rates of uptake, detoxification, and excretion (in Figure 4.1, [U], [D], and [E], respectively) regardless of total body burden.

A difficulty in relating metal uptake rates or tissue concentrations to toxicity has to do with the fact that organisms are complex systems consisting of many different physiological compartments. In addition, the size and the tendency of the bioreactive pool to be exceeded will differ among organisms depending on regulation,

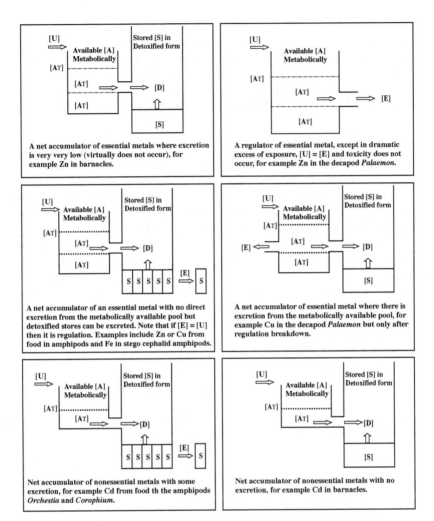

FIGURE 4.1 Theoretical schematic diagrams of uptake compartments for trace metals in marine organisms showing a pool of metabolically available metal, which can be physiologically regulated by balancing uptake with excretion and or detoxification. Toxic effects only occur when the rate of uptake exceeds the excretion or detoxification capacity and the maximum threshold for the level of metabolically available metal (i.e., the bioreactive pool) is exceeded. [A] includes the compartments or pools containing metabolically available metal — subcompartments or subpools consist of those required for essential functions and those containing excess. [A_R] is the pool within the metabolically available pool ([A]) that contains metal, fulfilling essential functions. [A_E] is the pool within the [A] pools that contains excess metal to cause effects if sufficiently elevated. [A_T] is the threshold level at which excess metabolically available metal causes effects. [U] is the uptake of metal, from the water column or via the gut. [D] is the detoxified metal, bound to ligands (e.g., but not limited to, metallothionein). [E] is the excretion of metal, by all mechanisms. [S] is stored metal, usually as granules. Note that the excess pool size may be very small relative to the required pool size, and, therefore, the total burden increase needed to produce effects may be a very small proportion of the total burden. (From Rainbow PS. 2002. Environ Pollut 120:497–507. With permission.)

detoxification, and excretion mechanisms. Thus, a total metal body concentration, specific tissue concentration, or uptake rate will only relate to metal toxicity if it reflects the interaction of the metal at the site of toxic action. Because uptake and elimination rates vary interspecifically, intraspecifically, and among tissues within a given organism, the exact mechanism of chronic metal toxicity will depend on the exposure scenario and may be difficult to ascertain under a given situation.

4.3.5 METAL EXPOSURE CONCENTRATIONS AND ACCUMULATION

On a whole-organism basis, bioaccumulation can be described by considering the organism as consisting of different kinetic compartments. These compartments may or may not reflect physiological units depending on the degree of detail in the model. In its most simple form, the organism is considered as 1 single box, with a single input for uptake and one output for excretion (e.g., similar to the top right panel in Figure 4.1). Although such a simple 1-compartment model is an oversimplification of reality, it can be a useful tool to describe the biodynamic relationship between exposure and accumulation, particularly if dietary and waterborne uptakes can be accounted for separately. Metal uptake in these biodynamic models is described by uptake rate constants (k_u) and excretion by an elimination rate constant (k_e). In the case of water exposure, the actual uptake rate is obtained by multiplying the uptake rate constant by the metal water concentration and the elimination rate by multiplying the body metal concentration by the elimination rate constant. Under steady-state conditions, uptake and elimination will balance, and the internal body concentration will remain constant. The uptake and elimination rate constants for metals are conditional constants that vary with the exposure conditions. However, k_u can vary with speciation, and some of the variability could be reduced if it were determined on the basis of free ion activity along with the concentrations and relative availability of other bioavailable metal species (Blust et al. 1992). The variability of uptake over metal exposure concentrations is illustrated by the kinetics of short-term metal uptake. These can be described by a Michealis–Menten-type transport model that characterizes the maximum tissue concentration (J_{max}) and the half-saturation constant, K_m, the metal exposure concentration at half of J_{max} (McDonald and Wood 1993; Simkiss and Taylor 1995; Van Ginneken et al. 1999; Wood 2001; Bury et al. 2003). These model variables fit a rectangular hyperbola curve characterized by a rapid increase that gradually levels off toward the maximum tissue concentration. In other words, initially the uptake rate constant is high, but then decreases as the transport system becomes saturated with increasing metal exposure concentration. The Michaelis–Menten-type transport model can also accommodate different types of interactions, such as competitive and other types of inhibition, which can alter the metal uptake rate constants (Blust 2001). In addition to short-term kinetics, metal uptake and elimination can vary with exposure, particularly in the context of chronic exposure. For example, responses to ongoing exposure can include a downregulation of uptake mechanisms and upregulation of elimination and detoxification mechanisms, particularly for essential elements for which body concentrations are regulated (Alsop et al. 1999; McGeer et al. 2000a, 2000b; Grosell et al. 2001), and in some instances, nonessential metals (Bury 2005). The consequence of having multiple factors that can influence uptake and elimination is that bioaccumulation is best modeled at equilibrium (so that uptake and elimination are relatively constant and balanced to give a consistent internal concentration). In turn, modeling at equilibrium requires some consideration of the physiological responses to metal exposure, for example, as characterized by the damage–repair model of McDonald and Wood (1993). The hypothesis of this model is that metal exposure disrupts existing homeostatic mechanisms (damage), which forces physiological adjustments (repair) that, if successful, result in the reestablishment of equilibrium but with different physiological constants (e.g., McGeer et al. 2000a, 2000b; Grosell et al. 2001). In terms of understanding and modeling bioaccumulation for the purposes of toxicity, one of the conceptual challenges is that, by definition, toxicity is associated with a disequilibrium condition.

4.4 LIMITATIONS OF CURRENT APPROACH TO BIOCONCENTRATION FACTORS (BCFs) AND BIOACCUMULATION FACTORS (BAFs)

4.4.1 METAL BIOACCUMULATION, TOXICITY, AND TROPHIC TRANSFER

One of the primary assumptions that makes BCF and BAF values suitable as indicators of bioaccumulation is that they are independent of exposure concentration (i.e., invariant uptake and elimination rate constants over a range of exposure concentrations). For neutral organic substances, this independence occurs because uptake is primarily via passive diffusion across the membrane lipid bilayer. However, inorganic substances have fundamental physicochemical differences compared to organic substances, and there is a complex relationship between metal bioaccumulation and exposure, especially across wide concentration ranges. Factors that could affect metal bioaccumulation include environmental conditions and biological factors, such as species-specific biodynamic considerations, essentiality, natural background, homeostasis, detoxification, and storage (although not all these are precisely defined nor is their influence precisely understood). The theoretical basis for applying BCF/BAF does not consider these complexities and, therefore, the validity of using BCF/BAF for the hazard classification or hazard assessment of metals is compromised as detailed in the following section.

4.4.1.1 Inverse Relationships

Inverse relationships occur between BCF or BAF and metal exposure concentration for essential and nonessential metals (McGeer et al. 2003). This not only complicates the theoretical aspect of using BCF/BAF values as an intrinsic property of a substance, but also results in elevated variability when data are compiled. Bioaccumulation of naturally occurring substances occurs along a continuum of exposure, and trace amounts of both essential and nonessential metals can be found in all biota (Cowgill 1976; Williams and Da Silva 2000). BCFs determined from natural conditions, which are characterized by low-exposure concentrations, can be as high as

300,000 and are generally meaningless in the context of evaluating potential for toxicity in relation to environmental hazard (McGeer et al. 2003). In addition, many aquatic organisms are also able to regulate internal metal concentrations through active regulation, storage, or combinations thereof (Adams et al. 2000; McGeer et al. 2003). Factors that influence metal uptake and bioaccumulation act at almost every level of abiotic and biotic complexity, including water geochemistry, membrane function, vascular and intercellular transfer mechanisms, and intracellular matrices. In addition, physiological processes (usually renal, biliary, or branchial) generally control elimination and detoxification processes. Storage adds additional controls on steady-state concentrations within the organism. Proportionally, less accumulation as exposure concentration increases means that there is an inverse relationship between exposure and metal BCFs and BAFs (McGeer et al. 2003). Further, when metal bioaccumulation is predominantly via mechanisms that demonstrate saturable uptake kinetics (note that some organic metal complexes can accumulate via diffusion; see first paragraph of Section 4.3.1), BCFs will decline at higher exposure concentrations.

4.4.1.2 Bioaccumulation in Relation to Chronic Toxicity

BCFs and BAFs are aggregate measures of all bioaccumulation processes and do not distinguish between different forms of bioaccumulated metal. The use of whole-organism metal concentrations for BCF and BAF calculations ignores the fact that internalized metals can occur in distinct pools, such as those involved in essential biochemical processes, those stored in chemically inert forms, and those with direct potential to bind at sites of toxic action (see Figure 4.1). The absence of a relationship between whole-body metal concentrations and toxic dose for many organisms complicates the application of BCFs and BAFs to metals. Such relationships are especially weak in organisms that use various mechanisms to store metals in detoxified forms, such as in inorganic granules (e.g., calcium phosphate-based, Cu–S complexes) or bound to metallothionein-like proteins. The use of granules is of particular importance in the context of BCFs, because extremely high body burdens are often associated with this storage mechanism and because this often (but not without exception) results in little or no toxicity to the accumulating organism or bioavailability to its predators. However, the relationship between accumulation and toxic effects is complex, and the protection afforded by detoxification mechanisms (for example, metallothionein, differences in granule compositions) can vary (Giguère et al. 2003). This relationship can also be complicated by the relative balance between the rates of metal uptake and detoxification that may lead to differing effects being associated with the same total body burden of metal (Rainbow 2002). Bioavailability of internal pools of bioaccumulated metal to consumers is also a factor that must be considered carefully, as this can vary according to the detoxification mechanism and digestive physiology of the consuming organism (see Section 4.3.2). To assess potential hazards associated with bioaccumulated metal, it would be necessary to distinguish between essential nutritional accumulation, benign accumulation (sequestering and storage), and accumulation that causes adverse chronic effects.

4.4.1.3 Trophic Transfer

Capturing the potential for metals to cause impacts via trophic transfer is one of the key goals associated with assessing metal bioaccumulation in the context of hazard evaluation. Because BCF calculations are based only on water concentrations, they do not consider dietary uptake, and, consequently, neglect the potential for impacts via that route. BAF values are calculated from water concentrations, and it is implicitly assumed that metal concentrations of field-collected organisms result from both waterborne and dietary exposures. It is also assumed that metal levels in an organism's diet result from the waterborne concentrations that it was exposed to. However, neither BCF nor BAF directly assess the potential for trophic transfer to result in toxicity. Although there are exceptions (for example, Se) and also specific circumstances where trophic transfer can be an issue, in general, documented occurrences of direct toxicity of diet-borne metals to consumer organisms have been limited to highly contaminated sites (Meyer 2005). Therefore, caution must be used in interpreting data on trophic transfer across single or multiple trophic levels as this is rare for inorganic metals. It can be confused with accumulation to meet physiological requirements (Rainbow 2002), and it may not even be a trophic-based phenomenon (Hare 1992). Effects of dietary exposure are metal-and species-specific, and, therefore, are most accurately assessed through studying specific food–consumer relationships.

4.4.2 IMPLICATION

In general, the use of BCFs and BAFs for metals as an indicator of chronic toxicity (both direct toxicity and trophic transfer impacts) is not supported by the current understanding of the science of metal uptake, distribution, and elimination. Any use of BCFs and BAFs should be done after data have been carefully evaluated and after the numerous scientific uncertainties have been investigated.

Bioaccumulation data for metals should generally not be used to estimate chronic toxicity, but when they are, this should be done with extreme caution. Instead, when the assessment end point is chronic toxicity, the use of chronic toxicity data is strongly preferred as the empirical demonstration of toxicity carries less uncertainty than a modeled estimate. Determining chronic toxicity should be relatively easy in some cases, such as direct waterborne toxicity, because for many metals, chronic data are available. However, novel approaches are needed to address the issue of the hazards associated with trophic transfer. The unit world model (UWM) offers one such novel approach to integrate both direct and trophic transfer, as well as chronic toxicity assessments into a unified assessment model (Chapter 3).

4.5 FURTHER GUIDANCE ON BIOACCUMULATION

4.5.1 BIODYNAMIC MODELS

Biodynamic models (Section 4.6.2.2), by their data demands, take into account both biology and geochemistry. Whether generic or site- and species-specific, uncertainty can be reduced to a far greater degree using biodynamic models as compared to generic BCFs or BAFs. Biodynamic models, or their more complex analogs, could be creatively used to constrain the bioaccumulative potential of a metal. Biodynamic-type models provide a preferable linkage to the UWM and a better basis for evaluating metals hazard (bioaccumulation in PBT) than do the empirical models, especially if the latter rely on generic constants. Most important, both geochemistry and biology add uncertainty to defining bioaccumulative potential.

However, because the use of biodynamic models requires a great deal of input information, although some regulatory frameworks require a generic approach in either hazard assessment or hazard ranking, other empirical models will be described as well for their use in a regulatory context.

4.5.2 Application of BCF and BAF Data

Recognizing that the UWM (Section 4.6) and the associated mechanistically-based bioaccumulation model proposed earlier will require additional development prior to their implementation for classifying and prioritizing metals, several interim alternatives for using bioaccumulation data in a hazard assessment context are considered and critiqued below. Within each of these suggestions, the broad question is whether or not the approach provides significant improvement over the current practice of using BCFs and BAFs in hazard assessment. More specifically, do the following interim alternatives:

1. Improve the linkage between bioaccumulation and direct chronic toxicity (i.e., to the bioaccumulating organism)?
2. Improve the ability to account for metal trophic transfer and the potential for secondary toxicity to predatory species?

4.5.2.1 Linking BCF with Chronic Lethality

Methodologies for linking chronic toxicity and BCFs would address one of the shortcomings of the current BCF application. Linkages between BCF and chronic toxicity can be done using mathematical relationships between body and water concentrations. This procedure has been applied using *Hyalella azteca*, an amphipod crustacean that is well suited for metal toxicology and bioaccumulation studies (Borgmann and Norwood 1995; MacLean et al. 1996; Borgmann et al. 2004). Body concentrations that occur at a chronic toxicity threshold (for example, the body concentrations associated with 25% mortality during 4 to 10 week exposure tests, or LBC_{25}s) can be relatively independent of exposure concentrations, indicating that metals must be accumulated by the organism to produce lethality, and that lethality occurs when tissue concentrations surpass a critical body concentration (CBC). For example, the concentration of Cd in water that caused 50% mortality in chronic toxicity tests was highly variable (> 35 fold), whereas Cd bioaccumulated in *H. azteca* during these same tests varied < 3 fold at the LC_{50} (Borgmann et al. 1991). Similar results have been shown for Tl and Ni with *H. azteca* (Borgmann et al. 1998, 2001). Furthermore, LBC_{25}s for nonessential, or sparingly essential metals such as Cd, Hg, Ni, Pb, and Tl are relatively constant (65 to 640 nmol/g dry weight), in

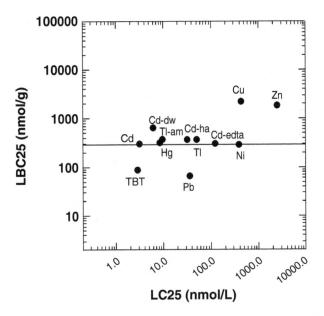

FIGURE 4.2 Relationship between the lethal body concentration causing 25% mortality in chronic toxicity tests with *H. azteca* (LBC$_{25}$, nmol/g dry weight) and the lethal concentration in water (LC$_{25}$) for various metals and TBT. Data for Cu and Zn have been corrected for background. (Data from Borgmann U. et al. 2004. Environ Pollut 131:469-484. With permission.) All data collected in tests using Lake Ontario water except where indicated as follows: am, artificial medium without K; dw, diluted with 90% distilled water; edta, 0.5 µM EDTA added; ha, 20 mg/l Aldrich humic acid added. The horizontal line is the geometric mean LBC$_{25}$, excluding Cu, Zn, and TBT (295 nmol/g).

spite of large differences in the waterborne concentrations that result in chronic toxicity (LC$_{25}$s, Figure 4.2). The LBC$_{25}$ for the organometal, TBT, is also similar. In contrast, the LBC$_{25}$s for Cu and Zn, which are essential metals required in numerous metabolic processes, are much higher (Figure 4.2).

Linking bioaccumulation data to chronic toxicity requires a measure of bioaccumulation that is independent of concentration. Borgmann et al. (2004) have shown that all metal bioaccumulation data collected to date for *H. azteca* could be fit to a rectangular hyperbola (see Section 4.3.5) of the form

$$C_{TB} = max \cdot C_W/(K + C_W) + C_{Bk} \tag{4.1}$$

which describes a hyperbolic increase to a maximum whole-body concentration as waterborne exposure concentration increases and where C_{TB} is total body metal concentration, max is the maximum whole-body concentration possible above background, C_W is the metal concentration in water, K is a constant representing the waterborne concentration at half of max, and C_{Bk} is the background metal concentration in the body. After fitting this equation to bioaccumulation data and deriving the max and K values, it is possible to calculate the ratio max/K. In some cases,

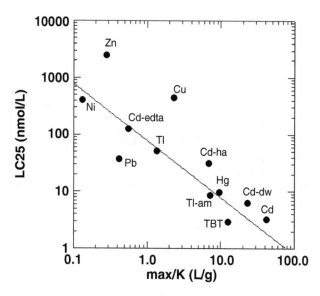

FIGURE 4.3 Relationship between the max/K (l/g wet weight) for metals and TBT in *H. azteca* and the lethal concentration in water (LC$_{25}$). Same data sources and symbols as in Figure 4.2. The line is the geometric mean best fit excluding Cu, Zn, and TBT (see chapter text) with a forced slope of 1.

bioaccumulation does not level off at high C$_W$ and max, and K cannot be estimated separately (e.g., Ni, Borgmann et al. 2004). In these cases, C$_W$ is much less than K at the range of metal concentrations investigated, and the above equation reduces to

$$C_{TB} = (max/K) \cdot C_W + C_{Bk} \qquad (4.2)$$

and the ratio max/K is estimated directly. The ratio of max/K is a background-corrected BCF extrapolated to a very low exposure concentration. Because it integrates data across concentrations, it can be considered to be independent of concentration, one of the problems associated with standard BCFs. This allows a comparison of bioaccumulation and chronic toxicity across metals (Figure 4.3, Table 4.1). For most metals, the log(max/K) values fall close to a line of slope −1, when plotted against log(LC$_{25}$). The essential metals Cu and Zn, however, have higher max/K values relative to the other metals, and therefore should not be included in comparisons using this methodology (Figure 4.3).

The max/K-based discrimination among the nonnutritional metals (Table 4.1) for waterborne LC$_{25}$ values arises because the LBC$_{25}$ (the LC$_{25}$ × BCF at the LC$_{25}$) values tend to be relatively constant (Figure 4.2) (Borgmann et al. 2004). It is important to note that max/K values for a given metal will vary with factors that alter the LC$_{25}$, for example, depending on water chemistry (see Cd and Tl in Figure 4.3). To illustrate the linkages between max/K and chronic toxicity, water-quality criteria and guidelines were compared to LC$_{25}$ and max/K values in Table 4.2. There is relatively good agreement between the criteria/guidelines and chronic

TABLE 4.1
Relationships between LC$_{25}$ and max/K for *H. azteca* in Comparison with Water Quality Guidelines (CCME), Water Quality Criteria (EPA), and Maximum Permissible Concentration in The Netherlands (NL-MPC)

Metal	max/K[a] (l/kg wet weight)*	LC$_{25}$[a] (μg/l)	CCME[b] (μg/l)	EPA[c] (μg/l)	NL-MPC[d] (μg/l)
Cd	42,200	0.36	0.017	0.25	0.4
TBT	12,700	0.34	0.008	0.063	0.014
Hg	9,650	1.95	0.1	0.77	0.2
Cu	2,360	28	2	9	1.5
Tl	1,380	10.5	0.8	na	1.6
Pb	424	7.6	2	2.5	11
Zn	287	165	30	120	9.4
Ni	133	23	65	52	5.1

Note: The LC$_{25}$ and max/K were measured in Lake Ontario water and the criteria/guideline values shown are correspondingly adjusted to a water hardness of 100 mg/l.

* l/g wet weight converted from dry weight using 0.19 g dry per 1.0 g wet.

Source: [a] Borgmann U. et al. 2004. Environ Pollut 131:469–484. [b] CCME (Canadian Council of Ministers of the Environment). 2002. Canadian water quality guidelines for the protection of aquatic life. Winnipeg, MB, Canada (calculated at 100 mg/l hardness). [c] USEPA 2002b. National Recommended Water Quality Criteria: 2002. EPA-822-R-02-047. Washington, D.C. (calculated at 100 mg/l hardness). [d] Crommentuijn T. et al. 2000. J Environ Manage 60:121–143; MPCs are based on the dissolved phase and include generic background concentrations for metals (except for TBT which are [in μg/l] for Cd, 0.08; for Hg, 0.01; for Tl, 0.04; for Pb, 0.2; for Cu, 0.4; for Zn, 2.8, and for Ni, 3.3).

LC$_{25}$ values for various metals and, as a result, an inverse relationship between these and max/K.

The max/K methodology achieves one of the objectives for considering bioaccumulation for hazard assessment. The approach, however, has several limitations, which include:

- No assessment of dietary toxicity.
- Limited number of metals with data.
- Relationship does not hold for nutritionally required and physiologically regulated metals (e.g., Cu and Zn).
- Exposure conditions affect the LC$_{25}$ determination, and thus subsequent ranking of metals.
- Representativeness of results from *H. azteca* to species that accumulate metals in detoxified forms, for example, granules, is unclear.

TABLE 4.2
Mean BCF/BAF and ACF Values for Selected Metals

Metal	Variable	Mean	Standard Deviation	CV (%)	N
Zinc	BCF: all data	3,394	8,216	242	133
	BCF: 10–110 µg/l	1,852	3,237	175	43
	ACF: all data	158	233	147	67
Cadmium	BCF: all data	1,866	4,844	260	226
	BCF: 0.1–3 µg/l	2,623	6,009	229	52
	ACF	352	615	175	96
Copper	BCF: all data	1,144	1720	150	122
	BCF: 1–10 µg/l	1,224	1,835	150	50
	ACF	456	659	145	46
Lead	BCF: all data	598	1,102	184	66
	BCF: 1–15 µg/l	410	647	158	14
	ACF	350	431	123	33
Nickel	BCF: all data	157	135	86	49
	BCF: 5–50 µg/l	106	53	50	27
	ACF	39	112	287	6
Silver	BCF: all data	1,233	2,338	190	29
	BCF: 0.4–5 µg/l	884	484	55	17
Mercury	BCF: all data	6,830	18,454	270	113
	BCF: 0.1–1 µg/l	10,558	23,553	223	54

Note: BCF values (including standard deviations and coefficients of variation) are provided over a limited exposure range that encompasses concentrations where chronic toxicity might be expected to begin occurring (based on water quality guidelines/criteria). (Adapted from McGeer JC. et al. 2003. Environ Toxicol Chem 22:1017–1037.) Insufficient data to calculate ACF values for Ag and Hg.

Further research is required to illustrate the robustness of this methodology for different metals, test species, and exposure conditions. Additionally, how the max/K values would be implemented in a regulatory context is unclear.

Other features of this methodology include the fact that whole-body burdens are used, and so there is no discrimination among toxic and other metal pools. This may be important as different exposure conditions may result in differences in uptake and may cause alterations in the relative pattern of metal accumulation within internal pools. This would cause variations in the body burden when threshold concentrations at the target site are finally reached (i.e., when toxicity occurs). Finally, measures that are designed as surrogates for chronic toxicity require the direct measurement of at least some chronic toxicity thresholds to validate the link between bioaccumulation and chronic toxicity. The empirical relationship that provides the link may introduce uncertainty, so direct measurement of chronic toxicity would be preferable for the purposes of hazard ranking. For *H. azteca*, chronic (minimum 4-week exposure) toxicity data are already available for a number of metals (Borgmann et al.

2004) and shorter-term (1-week) toxicity data are available for all metals and metalloids (Borgmann et al. 2005). Hence, the need for a surrogate measure of toxicity for this species is limited.

4.5.2.2 Accounting for Accumulation from Background Concentrations

As discussed previously, the existence of background metal concentrations in organisms (e.g., for normal metabolic requirements) can contribute to the observed inverse relationships between BCF and water concentrations, particularly when background concentrations are significant relative to newly accumulated metal. Thus, it stands to reason that separating the portion of metal that bioaccumulates from exposure under normal conditions from the portion that occurs as a result of exposure to elevated levels of metals may be one way to improve the linkage between exposure and toxicologically meaningful bioaccumulation. For instance, McGeer et al. (2003) adjusted metal concentrations in exposed organisms by subtracting metal concentrations in unexposed control organisms before calculating a value similar to the BCF. The accumulation factor (ACF) applies the concept behind the added risk approach proposed in the EU risk assessment process (for example, for Zn), accounting for the additional bioaccumulation that results from the added exposure.

This alternative has the conceptual advantage of addressing added accumulated metal explicitly, thereby separating the concept of essential or "normal" metal accumulation from the derivation of the BCF. The ACF would then represent the potential for accumulation above background levels in the organism. At least in some cases (that is, when significant metal regulation does not occur), this approach would reduce the impact of the inverse relationship on selecting the BCF. The disadvantage of this approach is that bioaccumulation is still not unambiguously linked to chronic toxicity. In addition, trophic transfer potential is not explicitly addressed. Also, in the context of risk screening or risk assessment, it may not be appropriate to subtract normal accumulation for evaluating the consequences of trophic transfer, given that background metals that have been accumulated and assimilated by prey organisms may be bioavailable to their predators.

4.5.2.3 Calculating BCF and BAF Values over a Limited Range of Concentrations

Another suggestion for limiting the effect of the inverse relationship on selecting BCF values for hazard assessments is to select BCFs (BAFs) that correspond to toxicologically relevant ranges in the environment (for example, near the applicable chronic water quality criterion). Conceptually, the advantage of this approach would be an improved linkage between the selected BCF (BAF) and the onset of chronic toxicity. This approach is therefore similar to the method described in Section 4.5.2.1 except that a range of species and exposure concentrations and conditions are considered. This measure was evaluated for some metals (McGeer et al. 2003), and this partial evaluation did not appear to substantially reduce the variability associated with BCF and BAF measurements across species (Table 4.2). Furthermore, the relationship between BCFs selected using this approach and chronic toxicity is compromised by the fact that water quality guidelines/criteria are influenced by responses of sensitive organisms, whereas BCF and BAF data for a metal are derived from a range of species including those that are more tolerant. In fact, the highest BCF/BAF values may be from the more insensitive organisms that use detoxification and storage mechanisms. Therefore, the selection of toxicologically relevant BCFs does not appear to reduce the uncertainties that are associated with the use of BCFs. Choosing BCFs from those organisms used to calculate toxicity thresholds may reduce uncertainty, but this again defeats the purpose of developing a surrogate, as direct measurement (i.e., chronic toxicity) is needed to develop relationships with the surrogate (i.e., BCF).

4.5.2.4 Bioaccumulation in Relation to Dietary Toxicity

Section 4.6 describes approaches for directly assessing the trophic transfer of metals and linking bioaccumulation to thresholds for dietary toxicity in wildlife. These approaches offer the obvious advantage of directly linking bioaccumulation potential to secondary toxicity via trophic transfer, which adds significantly to the interpretation of bioaccumulation data. Furthermore, following a careful review of the data, empirically based bioaccumulation relationships (that is, regressions of tissue concentrations vs. water concentrations) would appear to be available for incorporation into hazard assessment procedures in the near term without introducing substantial amounts of complexity. Such relationships inherently account for processes that contribute to the nonlinearity in bioaccumulation that is often observed for metals. The major disadvantage of these trophic transfer approaches is that they do not directly relate bioaccumulation to chronic toxicity experienced by the accumulating organism. However, this limitation can be overcome as understanding of the toxicological significance of metal residues in aquatic organisms grows.

4.6 INTEGRATION OF CHRONIC THRESHOLDS AND TROPHIC TRANSFER INTO THE UNIT WORLD MODEL

4.6.1 INTRODUCTION

The UWM approach (Chapter 3) begins with a model system to which metals are added until a relevant critical load is reached. One example of this approach would be to add metal to the model system and, after allowing equilibration into the various compartments, determining the amount of metal that can be added until the water concentration meets the chronic criterion for that metal. Equally, with the addition of metal to the model system it may be that critical concentrations will be exceeded in other compartments (for example, tissue burden and sediment) before waterborne guidelines/criteria are reached. From a hazard classification perspective, this approach allows for an integrated comparison of metals that is based on: (1) their geochemical properties, which determine the degree to which metals are distributed in various environmental phases (for example, dissolved in water, adsorbed to particles, incorporated into sediments, and so forth), and (2) the toxicity of metals to organisms relevant for each of these phases.

In addition to direct effects on aquatic biota via exposure to either water column or pore water metals, bioaccumulation and movement through the food web may cause adverse effects at concentrations below chronic criteria/guideline values. In these cases, this bioaccumulation will present the limiting hazard to the environment for some metals. Consequently, a food web submodel is required within the UWM to ensure that the environmental hazard of metals is not underestimated by ignoring this exposure pathway. The 2 specific goals of the submodel are to evaluate the degree to which different metals accumulate in aquatic organisms and to evaluate the biological consequences of this accumulation.

The end point of the food web submodel is an estimate of the metal concentrations within the tissues of a representative prey organism that result from a given waterborne metal concentration. These tissue concentrations would then serve as the exposure concentrations for upper trophic level predators. Although following the transfer of metals through natural food webs is complex, simplified approaches are available that are capable of estimating tissue concentrations from a few basic parameters. For example, biodynamic models can estimate steady-state tissue metal concentrations using dissolved and dietary metal concentrations as variable model parameters (Luoma and Rainbow 2005). Empirical approaches based on relationships between observed tissue concentrations and indices of bioaccumulation (e.g., BAF/BCF) are also available, but should not be carried forth into the UWM.

Several approaches are available for modeling metal bioaccumulation (Blust 2001; Paquin et al. 2003). For the purposes of incorporating a model that can provide estimates of tissue metal concentration in a prey organism, the ideal model would be based on quantitative measurements of bioaccumulation mechanisms (that is, uptake and elimination) and would account for all relevant metal uptake routes. For the present exercise and as an example of how bioaccumulation can be modeled effectively, a mechanistic biodynamic model was chosen to estimate tissue concentrations in the bivalve *Mytilus edulis*. The broad geographic distribution of *M. edulis*, the important role this bivalve plays in several marine and estuarine food webs, and the fact that bivalves are known to accumulate metals to relatively high concentrations compared with most other organisms, makes *M. edulis* a good indicator organism for this assessment. Equally importantly, the parameters needed to estimate tissue concentrations via the biodynamic model are available for this organism. The model is presented in Section 4.6.2.2.

4.6.2 TROPHIC TRANSFER MODELS

4.6.2.1 Conceptual Framework

The basic steps for incorporating bioaccumulation, as predicted from the biodynamic model, into the UWM are summarized as follows and depicted in Figure 4.4.

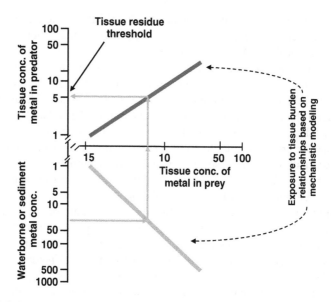

FIGURE 4.4 Conceptual framework for evaluating dietary toxicity potential using a bioaccumulation progression approach (accumulation from water or sediment to prey and then from prey to predator). (Based on Skorupa JP, Ohlendorf HM. 1991. In: Dinar A, Zilbeman D, editors. The economics and management of water and drainage in agriculture. Boston, MA: Kluwer, p. 345–368. With permission.) Note that some mechanistic models, for example biodynamic models, incorporate waterborne and dietary bioaccumulation simultaneously.

1. The water quality criterion/guideline or, if it is more appropriate for the food web being studied, the sediment criterion/guideline is selected for model input.
2. Using this exposure concentration, a tissue concentration for the selected aquatic organism can be estimated using the modeling approaches described in detail below.
3. The resulting predicted tissue concentration can then be compared to a dietary threshold for the selected consumer organism.
4. If the predicted metal tissue concentration in the predator organism exceeds its tissue burden threshold, then bioaccumulation becomes the critical pathway and biodynamic modeling parameters should be incorporated into the UWM.

Within the UWM framework, if the predicted tissue concentration in the prey organism at the water quality criterion/guideline is less than the dietary threshold for the consumer organism, then dietary toxicity does not represent the limiting pathway with respect to environmental hazard; rather, the overall hazard of the substance will be determined by toxicity thresholds based on direct toxicity to aquatic life. On the other hand, if the predicted tissue concentration in the prey organism at the water quality criterion/guideline exceeds the dietary threshold for the consumer organism, then dietary toxicity is the limiting pathway and a back calculation to the appropriate safe concentration in water or sediment must be made for use in the UWM framework.

In the next subsection, specific details on the modeling approaches for predicting tissue concentrations in prey organisms are described.

4.6.2.2 Biodynamic Bioaccumulation Models

The biodynamic model was developed to predict steady-state tissue concentrations (C_{ss}) in aquatic organisms based on integrated metal accumulation from waterborne and diet-borne uptake routes (Schlekat et al. 2001; Luoma and Rainbow 2005):

$$C_{ss} = (k_u \times C_W)/(k_{ew}) + (AE \times IR \times C_F)/(k_{ef}) \qquad (4.3)$$

where k_u = dissolved metal uptake rate constant (l/g/d), C_W = dissolved metal concentration (μg/l), AE = assimilation efficiency (%), IR = ingestion rate (mg/g/d), C_F = metal concentration in food (for example, phytoplankton, suspended particulate matter, and sediment) (μg/g), and k_{ew} and k_{ef} = efflux rates from waterborne and diet-borne metal, respectively (l/d). Water and food concentrations, C_W and C_F, can be site-specific in nature, or they can be conceptual for illustrative purposes. The other model components (e.g., AE, IR, and k_u) are species-specific physiological constants that are determined in the laboratory. Effects of individual components and their interactions have been the focus of several reviews (e.g., Reinfelder et al. 1998; Wang and Fisher 1999).

Model predictions have agreed well with field-measured metal concentrations in several studies that have covered a wide range of organisms, as well as a diversity of food webs, habitats, food types, and metals (Reinfelder et al. 1998). For example, Griscom et al. (2002) modeled the accumulation of Ag, Cd, and Co by the bivalve *Macoma balthica* from surficial sediments in San Francisco Bay. Mean predicted concentrations of Ag and Cd were 6.3 and 0.2 μg g^{-1}, respectively, whereas mean measured concentrations were 7.6 and 0.3 μg g^{-1}, respectively. As long as the values of C_W and C_F that are to be modeled are within the metal concentrations used to measure physiological uptake parameters in the laboratory, model predictions have been shown to be accurate. In a recent review, Luoma and Rainbow (2005) compared biodynamic model predictions with independent measured tissue concentrations obtained from field studies. The data set consisted of 15 separate studies that included comparisons of 7 different metals and 14 different species. A strong relationship (r^2 = 0.98) was observed between predicted and observed tissue concentrations, further supporting the validity of the biodynamic model approach.

Other mechanistic models are available to estimate tissue metal concentrations. For example, Clason et al. (2004) and Kahle and Zauke (2003) have developed bioaccumulation models for amphipod crustaceans. These 2-compartment models are based on tissue concentration data arising from exposures to dissolved metals and utilize statistical analyses to derive uptake and elimination rate constants. These constants are used to estimate tissue concentrations in a model where dissolved metal concentrations and background tissue concentrations are the other variables.

These models show good agreement between predicted and observed concentrations for some of the metals studied.

In the 2-compartment models, experiments with increasing levels of metals in the exposure medium (dissolved metal concentrations) have been associated with decreases in metal uptake rate constants, and these have been attributed to saturation of uptake kinetics (Clason et al. 2004). Therefore, tissue concentrations from these models could be estimated in a concentration-dependent way if needed. Biodynamic models have not accounted for possible effects of saturation of uptake kinetics because it is assumed that the uptake rate constant k_u describes a linear increase in uptake with exposure concentration. This assumption is supported by results of laboratory experiments showing linear increases of metal uptake rates with exposure concentrations that are much higher than those occurring in nature (Luoma and Rainbow 2005). For example, Wang et al. (1996) observed that the dissolved uptake rate for *M. edulis* increased linearly with increased dissolved metal concentrations over wide concentration ranges, for example, 0.1 to 10 μg/l for Cd and 0.5 to 300 μg/l for Zn. These ranges bracket environmentally relevant concentrations and extend beyond current environmental quality guidelines (for example, the EPA water quality criterion for Cd is 0.25 μg/l). Clason et al. (2004), on the other hand, reported a decrease in k_u with increasing dissolved Cd concentration for the amphipod *Gammarus oceanicus*, suggesting a saturation of metal uptake mechanisms. However, the exposure concentrations used by Clason et al. (2004) ranged from 5 to 30 μg Cd/l, which is well above both reported ambient concentrations and environmental quality guidelines. Therefore, saturation of uptake, and the potential for this phenomenon to influence bioaccumulation predictions, needs to be considered on a case-by-case basis. In the case of *M. edulis*, it does not appear that saturation of uptake is a factor that needs to be considered for the use of the biodynamic model.

It is possible to derive biodynamic models that incorporate variable metal uptake and elimination rates (see Blust 2001 for a review). For example, in terms of Michaelis–Menten-type kinetics the uptake rate constant equals $k_u = J_{max}/(K_m + C_{exp})$. Depending on the exposure conditions (that is, bioavailability) and physiology of the organism, both J_{max} and K_m may change and hence also k_u. For example, the effect of calcium on cadmium uptake can be easily incorporated into the biodynamic model by incorporating a competitive effect in the Michaelis–Menten model. In practice, this means that the K_m value will increase (lower affinity) with increasing calcium (e.g., Chowdhury and Blust 2001). Effects of chemical speciation can be incorporated by replacing the C_w by the activity of the metal species available for uptake. Other effects can be dealt with in a more or less similar manner. However, it may not be possible to incorporate the effects of, for example, salinity and temperature on k_u through speciation alone. These effects can also be incorporated into the biodynamic model by Michaelis–Menten or related approaches. In practice, the physical and chemical conditions within the UWM should be standardized to avoid the influence of exposure conditions and organism physiology on bioaccumulation estimates.

One main limitation of the aforementioned two-compartment bioaccumulation models is that they do not account for contributions of diet-borne exposure in the estimation of tissue metal concentrations. The relative importance of diet-borne metal varies across metals and organisms, but is a major contributor to total tissue metal concentration under most circumstances (Schlekat et al. 2001). For example, Wang et al. (1996) showed that the proportion of diet-borne metal uptake for *M. edulis* increased from Cd (24 to 47%, depending on food source) to Ag (43 to 69%) to Zn (48 to 67%) to Se (> 96%). Failure to account for diet-borne metal exposure by a prey organism will underestimate diet-borne exposure to a predator, which is the focus of the conceptual model discussed earlier. Therefore, selection of the biodynamic modeling approach is appropriate as it accounts for diet-borne exposure.

Until recently, biodynamic models were available only for those metals with gamma-emitting radioisotopes, and particularly those isotopes that are relatively long lived. This is because the protocols for determining uptake rate constants were based on short exposures that resulted in low metal concentrations and necessitated the low detection limits offered by radioisotopes. Often, the accumulations of metal could not be distinguished from existing background concentrations. Also, the protocols for determining assimilation efficiency from food required repeated nondestructive analysis of organisms. Recent developments and new analytical methods have removed these limitations. Work by Croteau et al. (2004) and Evans et al. (2002) used stable isotopes of Cu and Cd, respectively, to determine uptake and other dynamic properties. Croteau et al. (2004) were successful in determining uptake and elimination rate constants for the bivalve *Corbicula fluminea* using ^{65}Cu. Their work represents the first protocol for determining biodynamic model parameters with stable isotopes.

For the example presented below, the physiological parameters for the biodynamic model for *M. edulis* were taken from Wang and Fisher (1996). This study determined uptake and elimination kinetics for Ag, Cd, Se, and Zn, among other metals. Dissolved concentrations ranged from ambient background concentrations to reasonable worst-case concentrations, for example, chronic values from Canada and the United States. The biodynamic model requires concentrations of metals within the food of *M. edulis*, and these were estimated using the distribution coefficient, K_D (l/kg) and the following formula: $C_F = C_w \times K_D$. The use of a single K_D value to estimate diet-borne exposure has drawbacks. First, it ensures that estimated tissue concentrations will increase linearly with dissolved concentrations. This brings the same problems that were associated with the use of BCF/BAF and limits the utility of applying the model over a range of dissolved concentrations, as the relative ranking of tissue concentrations will remain independent of dissolved concentration. To be representative of what occurs in a given natural system, site-specific geochemical parameters (e.g., measured concentrations in the dissolved phase and in relevant food sources) should be used. Although this may be important in terms of site-specific risk assessment, the issue of representativeness may not be relevant for the current hazard classification exercise, the goal of which is to determine when tissue concentrations in a generic prey organism exceed tissue reference values for a generic predator. For the present exercise, average K_Ds measured for San Francisco Bay were used to be consistent with Wang et al. (1996). A more objective approach would be to use the geometric mean of K_Ds from relevant habitats of *M. edulis*, for example, temperate coastal regions. In the context of hazard assessment and the current state of the science of biodynamic modeling, the use of K_D is an important issue that awaits further refinement to account for the variation and differences among species and conditions.

This model assumes that mussels are feeding on seston, which is composed of suspended algae and organic and inorganic particulate material. The efficiency with which organisms assimilate metals from food can vary tremendously, and is a function of physiological and geochemical factors (Reinfelder et al. 1998). For example, assimilation efficiencies (AE) for *M. edulis* from seston can vary considerably depending on the nature of the suspended particulate matter (e.g., AE for Se can vary from 30 to 70%); therefore, the mean of AEs reported by Wang and Fisher (1996) was used.

4.6.2.3 Use of Model Outputs

Example outputs using the 2 modeling approaches are presented below. For demonstration purposes we derived predicted tissue concentrations in bivalves via both water-column and sediment pathways. The starting water concentrations were chronic water quality values for the United States and Canada. Corresponding sediment concentrations were derived through use of a simple K_D. Note that estimation of sediment concentrations using this approach ignores a number of important processes (for example, burial, sulfide portioning, and resuspension) that will be incorporated in the UWM. Results from each model were then compared to dietary thresholds for aquatic-dependent wildlife (Table 4.3).

Estimated tissue concentrations appear to be improbably high when U.S. water quality criteria are used as the input. For example, predicted Zn tissue concentrations for *M. edulis* were 15,000 µg Zn/g when the dissolved concentration was 120 µg Zn/l. Databases of bivalve concentrations suggest that *M. edulis* is unlikely to achieve this concentration in nature, if only because *M. edulis* can partially regulate internal Zn concentrations (Philips and Rainbow 1993; Wang 2002). Concentrations of Zn as high as 23,300 µg/g (dry weight) have been measured in barnacles (Rainbow and Blackmore 2001), although Zn concentrations in mussels that occur in the same habitat are typically 1 to 2 orders of magnitude lower (Philips and Rainbow 1988). On the other hand, it may be questioned whether or not *M. edulis* would occur in areas that showed permanent exceedences of EPA water quality criteria (WQC). A further consideration would be that the biodynamic model needs further refinement to be applied to Zn. As shown in other work (Borgmann et al. 2004), nutritionally required elements can be closely regulated, for example, via active control of uptake and elimination.

The important step in the conceptual application of model outputs within the UWM is to compare model outputs with dietary threshold values. Predicted tissue concentrations exceeded dietary thresholds (i.e., toxicity reference values [TRVs]) for Se and Zn regardless of the regional water quality guideline/criterion used as the starting point (Table 4.3). Taken at face value, this would indicate that dietary toxicity represents the limiting pathway with respect to environmental hazard for these metals. For other metals, for example Cd, the overall hazard will be determined by toxicity thresholds based on direct toxicity. This demonstrates that it is possible to discriminate among the extent to which different metals bioaccumulate and also

TABLE 4.3
Comparison of Predicted Tissue Concentrations from Biodynamic
and Empirical Modeling to Dietary Threshold Values for Aquatic
Dependent Wildlife

Metal	Dissolved Exposure	K_D	Predicted Food Concentration ($\mu g/g$)	Predicted Tissue Concentration ($\mu g/g$)	Dietary Threshold ($\mu g/g$)
Ag	0.003[a]	150000	0.39	0.49	n/a
Ag	0.1[b]	150000	15	19.0	n/a
Ag	3[c]	150000	450	569	n/a
Cd	0.01[a]	5000	0.48	0.48	45[d]
Cd	0.03[b]	5000	0.15	1.4	45[d]
Cd	0.25[c]	5000	1.25	11.9	45[d]
Cr	2[b]	1000	2	1.1	22[e]
Cr	11[c]	1000	11	5.6	22[e]
Se	0.025[a]	3000	0.25	1.6	5.6[f]
Se	1[b]	3000	3	62.7	5.6[f]
Se	5[c]	3000	15	313.6	5.6[f]
Zn	0.32[a]	20000	6.3	41.4	177[g]
Zn	30[b]	20000	600	3942	177[g]
Zn	120[c]	20000	2400	15768	177[g]

Source: [a] Wang W-X. et al. 1996. Mar Ecol Progr Ser 140:91–113; (lower range of concentrations measured for San Francisco Bay). [b] CCME (Canadian Council of Ministers of the Environment). 2002. Canadian water quality guidelines for the protection of aquatic life. Winnipeg, MB, Canada. [c] USEPA. 2002b. National Recommended Water Quality Criteria: 2002. EPA-822-R-02-047. Washington, D.C. [d] Wilson RH. et al. 1941. J Pharmacol Exp Ther 71:222-235. [e] Haseltine SD. et al. 1985. As cited in Sample BE. et al. 1996. Toxicological benchmarks for wildlife: 1996 revision. Oak Ridge National Laboratory, ES/ER/TM-86/R3. Oak Ridge, TN: Oak Ridge National Laboratory. [f] Heinz GH. et al. 1989. J Wildl Manage 53:418–428. [g] Hamilton RP. et al. 1979. J Food Sci 44: 738–741.

to discriminate based on the biological consequences of this bioaccumulation. A ranking system based on comparison of predicted tissue concentrations to wildlife/aquatic TRVs is therefore feasible; however, it must be stressed that many steps need to be taken before this approach can be applied for regulatory purposes, because of uncertainties in making the link between dissolved metal concentrations to concentrations in food items, and in the validity of currently available TRVs.

4.6.3 UNCERTAINTIES

4.6.3.1 Bioaccumulation Models

It is important to demonstrate that the models used in the bioaccumulation module represent what occurs in nature. To this end, it is important to evaluate the validity of the biodynamic model. One area of possible concern is that the estimated tissue concentrations at high dissolved concentrations appear to be improbably high. Luoma and Rainbow (2005) showed that biodynamic model predictions were highly predictive of observed tissue concentrations and that this relationship held across a wide range of metals, organisms, and habitats. Judging the validity of the model for the present exercise is not possible because *M. edulis* does not occur in areas where dissolved concentrations are continuously at EPA WQC levels, and the comparable observed tissue concentrations are therefore not available to compare with model estimates. Given that there is no way to conclude that the biodynamic model estimates were unrealistically high, the factors affecting model estimates can only be discussed conceptually.

One set of reasonable possibilities relates to the conditions under which the kinetic parameters for *M. edulis* were determined. That is, it is possible that the range of conditions used in the present modeling exercise went beyond those used to develop the model. It was assumed that parameters such as uptake kinetics, assimilation efficiency, and elimination kinetics remained constant over the concentration ranges evaluated in the present exercise. Wang and Fisher (1996) observed a linear relationship between dissolved uptake kinetics and dissolved metal concentrations across a wide concentration range (e.g., from 0.5 to 300 µg/l for Zn); therefore, the possibility that uptake kinetics were saturated at the higher dissolved concentrations does not appear to be a likely explanation for the high estimated tissue concentrations. However, the uptake experiments conducted by Wang and Fisher (1996) employed naïve mussels (i.e., mussels acclimated to background metal concentrations) and short (2-hour) exposure periods. It is possible that mussels acclimated to higher metal concentrations would have developed the means to decrease influx rates or increase efflux rates. Investigating these possibilities is a clear research need.

Other physiological processes may also be dependent on exposure concentrations. For example, uptake from the dietary pathway could be affected at high concentrations of metal in food by decreased ingestion rates or saturation of uptake kinetics across the gut. Wang and Fisher (1996) took pains to prepare experimental food such that the metal concentrations within the food were representative of what occurred in the natural systems of interest. In other words, the possibility that dietary uptake could be affected by metal concentration was not addressed in these experiments. Experimental evidence suggests that assimilation efficiency does decrease at very high metal concentrations, at least for some metal/organism combinations (Schlekat et al. 1999). Elimination rates, on the other hand, are less likely to explain why estimated tissue concentrations appeared so high. Elimination rates are a function of tissue concentration and the rate constant of loss (k_e), and although the rate of loss will increase as tissue concentrations increase, there is no evidence to show that k_e will increase based on exposure concentration.

To utilize the mechanistic bioaccumulation model for the purposes of hazard assessment within the UWM, kinetics for metal uptake and elimination must be developed for the model organism at the dissolved concentrations that are equivalent to the chronic toxicity values, that is, the alternative critical load outcomes. Experiments to obtain these data should be relatively straightforward, and basically represent a modification of existing methodologies.

Another issue that needs to be considered is the choice of model organism. The ideal organism should be widely distributed in nature, be an important and representative prey organism, and should represent a reasonable worse case with respect to metal accumulation. That is, the model organism should not discriminate in the degree to which it accumulates metals. The present exercise utilized *M. edulis.* Although the model predicted high Zn concentrations at the EPA WQC, *M. edulis* is known to regulate Zn tissue concentrations. Thus, it is likely that the model output does not reflect what occurs in nature and that, at least for Zn, *M. edulis* does not reflect a reasonable worse case. Therefore, other organisms should be considered. Reinfelder et al. (1997) showed that *M. edulis* has significantly lower trophic transfer potential for Ag, Cd, and Zn than the oyster, *Crassostrea virginica.* Trophic transfer potentials for Se were relatively similar. This suggests that the use of *C. virginica* might be more relevant for this purpose.

Factors other than kinetics may also contribute to the departure from biodynamic modeling estimates. For example, it is possible that the K_Ds used overestimated the predicted food concentrations, and this, in turn, overestimated the dietary contribution to the steady-state tissue concentration. Metal partitioning to particles varies according to many variables (Turner 1996), and it is possible that metal partitioning would decrease with increasing metal concentrations, in which case the use of K_Ds that were developed at background concentrations would be inappropriate to estimate partitioning when dissolved metal concentrations were at the EPA WQC. It is essential that dietary exposure, at K_Ds or food concentrations expected from nature, be included in these chronic toxicity tests.

4.6.3.2 Toxicity Reference Values (TRVs)

No matter which model is used, an important issue to consider in this process is the derivation of dietary toxicity threshold. The threshold concentrations used in this exercise were derived from the literature and represent the geometric mean of the NOEC and LOEC for the most sensitive aquatic-dependent species tested. Although this approach results in what we believe to be realistic thresholds, alternative methods for deriving dietary toxicity thresholds can result in substantially lower values. For example, the EPA (1999b) derived a threshold concentration for zinc from mouse experiments. Using allometric relationships, the mouse threshold concentration was extrapolated to a raccoon, which represents a potential consumer of bivalves. As such, the *M. edulis* to raccoon pathway is ecologically relevant. However, using this approach, a threshold concentration for raccoons of 55 mg Zn/kg food is estimated. This is problematic as it is below tissue concentrations needed to meet metabolic requirements for bivalves (White and Rainbow 1985) and below tissue concentrations in bivalves collected from pristine areas with no metal contamination. Thus, it is clear that derivation of environmentally realistic dietary toxicity thresholds is critical to the successful application of the proposed approach.

4.6.3.3 Protectiveness of Environmental Quality Standards

Finally, it is possible that the ambient water quality criteria/guidelines for some metals (Zn and Se may be examples) could underprotect aquatic life under the scenarios presented. One of the values of a model is to raise such unexpected questions. The Zn values were derived from experiments that do not include dietary exposure. Although chronic criteria/guidelines do involve long-term chronic toxicity tests, those exposures can be different compared to exposures in natural waters. The correction for these uncertainties is arbitrary. It could be argued that the uncertainties created in those experiments are as great as the uncertainties in the bioaccumulation model. As noted above, for the model to be valid, and thus the water quality guideline/criteria values to be flawed, several conditions are necessary. The K_Ds must prove to be reasonable at these high concentrations in nature (that is not the case for many Se-contaminated environments, as processes other than absorption are responsible for transforming dissolved Se into particulate forms; but it may be for other metals). Uptake rates from dissolved and dietary sources must increase linearly with concentration and, if they do not, another approach such as Michaelis–Menten kinetics must be used to predict steady-state uptake. Similarly, rate constants for loss can be incorporated under the assumption of linearity; if evidence arises illustrating otherwise, then the model should be adjusted to account for this. In any case, the assumptions applied to the model should be clearly discussed and validated.

Presuming the physiological and physiochemical assumptions are correct, it is reasonable to expect that mussels would die if very rapid uptake rates overcome their ability to regulate. Directly testing the model forecasts would require finding such conditions in nature where mussels are alive, or creating such conditions in an experiment that simulates nature (including a K_D typical of nature, relevant dietary exposure, and time of exposure up to 200 d). Testing the assumptions is another approach, as suggested above. These are important points to consider for the evaluation of hazard assessment within the UWM framework; in order for the framework to be accepted as being reliable and credible, the assumptions upon which the model is based ultimately need to be validated.

4.7 CONCLUSIONS

The mechanisms by which metals are taken up and distributed within organisms, how they reach target sites, and how impairment occurs are the key features of bioaccumulation that need to be understood in order to develop links between exposure and the potential for chronic impacts. These aspects of bioaccumulation need to be understood in the context of 2 routes of exposure, waterborne and dietary. Although bioaccumulation of substances is of potential concern, simplistic measurements such as bioconcentration factors (BCFs) and bioaccumulation factors (BAFs) are not reliable as generalized surrogates for chronic toxicity of metals, nor are they useful in evaluating dietary effects to predators consuming these organisms with perhaps the exception of specific, constrained, and well-characterized predator–prey interactions. The reasons for this unreliability arise from the complexities of metal bioaccumulation, which include: saturation kinetics for uptake; homeostatic mechanisms to control internal burden; natural background accumulation; the essentiality of some metals; and the ability to detoxify, store, and excrete excess accumulated metal.

Improvements are required in the use of bioaccumulation in environmental assessments. There are some potential approaches for dealing with the inadequacies of BCF and BAF but, overall, links between bioaccumulation and direct chronic toxicity are still ambiguous. Tools to assess the toxic consequences of bioaccumulation to secondary consumers need to be developed. There are bioaccumulation models (both mechanical and empirical) that show promise in this regard. Among the most important data needs for trophic transfer bioaccumulation models are thresholds for dietary toxicity. In the meantime, and for the foreseeable future, whenever possible, assessments of chronic toxicity should use chronic toxicity data (which are available for many metals) with an emphasis on studies that account for dietary exposures rather than a surrogate such as bioaccumulation.

Incorporation of bioaccumulation into the UWM can be achieved by using submodels that account for dietary impacts of metal bioaccumulation to secondary consumers (predators); biodynamic modeling approaches to predict bioaccumulation show promise in this regard. One methodological approach would be to create a generic food web that estimates tissue metal concentrations in a cosmopolitan prey organism, for example, a bivalve, from dissolved and diet-borne exposure routes. Estimated tissue concentrations would then be compared to dietary toxicity thresholds. Development of these models depends on establishing ecologically relevant tissue threshold concentrations for metals.

REFERENCES

Adams WJ, Conard B, Ethier G, Brix KV, Paquin PR, Di Toro DM. 2000. The challenges of hazard identification and classification of insoluble metals and metal substances for the aquatic environment. Human Ecol Risk Assess 6:1019–1038.

Alsop D, McGeer JC, McDonald DG, Wood CM. 1999. Costs of chronic waterborne zinc exposure and the consequences of zinc acclimation on gill/zinc interactions of rainbow trout in hard and soft water. Environ Toxicol Chem 18:1014–1025.

Blust R. 2001. Radionuclide accumulation in freshwater organisms: concepts and models. In: Van der Stricht E, Kirchmann R, editors. Radioecology, radioactivity and ecosystems. Liège, Belgium: International Union of Radioecology, p. 57–89.

Blust R, Kockelbergh E, Baillieul M. 1992. Effect of salinity on the uptake of cadmium by the brine shrimp *Artemia franciscana*. Mar Ecol Progr Ser 84:245–254.

Borgmann U, Norwood WP. 1995. Kinetics of excess (above background) copper and zinc in *Hyalella azteca* and their relationship to chronic toxicity. Can J Fish Aquat Sci 52:864–874.

Borgmann U, Norwood WP, Babirad IM. 1991. Relationship between chronic toxicity and bioaccumulation of cadmium in *Hyalella azteca*. Can J Fish Aquat Sci 48:1055–1060.

Borgmann U, Cheam V, Norwood WP, Lechner J. 1998. Toxicity and bioaccumulation of thallium in *Hyalella azteca*, with comparison to other metals and prediction of environmental impact. Environ Pollut 99:105–114.

Borgmann U, Neron R, Norwood WP. 2001. Quantification of bioavailable nickel in sediments and toxic thresholds to *Hyalella azteca*. Environ Pollut 111:189–198.

Borgmann U, Norwood WP, Dixon DG. 2004. Re-evaluation of metal bioaccumulation and chronic toxicity in *Hyalella azteca* using saturation curves and the biotic ligand model. Environ Pollut 131:469–484.

Borgmann U, Couillard Y, Doyle P, Dixon GD. 2005. Toxicity of sixty-three metals and metalloids to *Hyalella azteca* at two levels of water hardness. Environ Toxicol Chem 24:641–652.

Bury NR. 2005. The changes to apical silver membrane uptake, and basolateral membrane silver export in the gills of rainbow trout (*Oncorhynchus mykiss*) on exposure to sublethal silver concentrations. Aquat Toxicol 72:135–145.

Bury NR, Walker PA, Glover CN. 2003. Nutritive metal uptake in teleost fish. J Exp Biol 203:11–23.

Campbell PGC. 1995. Interactions between trace metals and aquatic organisms: a critique of the free-ion activity model. In: Tessier A, Turner A, editors. Metal speciation and bioavailability in aquatic systems. Chichester, UK: John Wiley and Sons, p. 45–102.

Campbell PGC, Clearwater SJ, Brown PB, Fisher NS, Hogstrand C, Lopez GR, Mayer LM, Meyer JS. 2006. Digestive physiology, chemistry and nutrition. In: Meyer JS, editor. Diet-borne metal toxicity to aquatic organisms. Pensacola, FL: SETAC Press (in press).

CCME (Canadian Council of Ministers of the Environment). 2002. Canadian water quality guidelines for the protection of aquatic life. Winnipeg, MB, Canada.

CEPA (Canadian Environmental Protection Act). 1999. Ottawa, ON, Canada.

Chen Z, Mayer LM, Weston DP, Bock MJ, Jumars PA. 2002. Inhibition of digestive enzyme activity by copper in the guts of various marine benthic invertebrates. Environ Toxicol Chem 21:1243–1248.

Cheung M, Wang WX. 2005. Influence of subcellular compartmentalization in different prey on the transfer of metals to a predatory gastropod. Mar Ecol Prog Ser 286:155–166.

Chowdhury MJ, Blust R. 2001. A mechanistic model for the uptake of waterborne strontium in the common carp (*Cyprinus carpio* L.). Environ Sci Technol 35:669–675.

Clason B, Gulliksen B, Zauke GP. 2004. Assessment of two-compartment models as predictive tools for the bioaccumulation of trace metals in the amphipod *Gammarus oceanicus*. Segerstrale, 1947, from Gunnfjord (Northern Norway). Sci Tot Environ 323:227–241.

Collins JF, Franck CA, Howdley KV, Ghishan FK. 2005. Idenitfication of differentially expressed genes in response to dietary iron deprivation in rat duodenum. Am J Physiol 288:G964–G971.

Cowgill UM. 1976. The chemical composition of two species of *Daphnia*, their algal food and their environment. Sci Tot Environ 6:79–102.

Crommentuijn T, Sijm D, De Bruijn J, Van den Hoop M, Van Leeuwen K, Van de Plassche E. 2000. Maximum permissible and negligible concentrations for metals and metal-loids in The Netherlands, taking into account background concentrations. J Environ Manage 60:121–143.

Croteau M-N, Luoma SN, Topping BR, Lopez CB. 2004. Stable metal isotopes reveal copper accumulation and loss dynamics in the freshwater bivalve *Corbicula*. Environ Sci Technol 38:5002–5009.

Evans RD, Balch GS, Evans HE, Welbourne PM. 2002. Simultaneous measurement of uptake and elimination of cadmium by caddisfly (*Trichoptera*) larvae using stable isotope tracers. Environ Toxicol Chem 21:1032–1039.

Finney LA, O'Halloran TV. 2003. Transition metal speciation in the cell: insights from the chemistry of metal ion receptors. Science 300:931-936.

Giguère A, Couillard Y, Campbell PGC, Perceval O, Hare L, Pinel-Alloul B, Pellerin J. 2003. Steady-state distribution of metals among metallothionein and other cytosolic ligands and links to cytotoxicity in bivalves living along a polymetallic gradient. Aquat Toxicol 64:185–203.

Glover CN, Bury NR, Hogstrand C. 2003. Zinc uptake across the apical membrane of freshwater rainbow trout intestine is mediated by high affinity, low affinity and histidine-facilitated pathways. Biochim Biophys Acta 1614:211–219.

Griscom SB, Fisher NS, Luoma SN. 2002. Kinetic modelling of Ag, Cd and Co bioaccumulation in the clam *Macoma balthica*: quantifying dietary and dissolved sources. Mar Ecol Prog Ser 240:127–141.

Grosell M, McGeer JC, Wood CM. 2001. Plasma copper clearance and biliary copper excretion are stimulated in copper-acclimated trout. Am J Physiol 280:R796–R806.

Hamilton RP, Fox MRS, Fry BVE, Jones AOL, Jacobs RM. 1979. Zinc interference with copper, iron and manganese in young Japanese quail. J Food Sci 44: 738–741.

Hare L. 1992. Aquatic insects and trace metals: bioavailability, bioaccumulation and toxicity. Crit Rev Toxicol 22:327–369.

Haseltine SD, Sileo L, Hoffman DJ, Mulhern BD. 1985. Effects of chromium on reproduction and growth in black ducks. As cited in Sample BE, Opresko DM, Suter GW. 1996. Toxicological benchmarks for wildlife: 1996 revision. Oak Ridge National Laboratory, ES/ER/TM-86/R3. Oak Ridge, TN: Oak Ridge National Laboratory.

Heinz GH, Hoffman DJ, Gold LG. 1989. Impaired reproduction of mallards fed an organic form of selenium. J Wildl Manage 53:418–428.

Huffman DL, O'Halloran TV. 2001. Function, structure, and mechanism of intracellular copper trafficking proteins. Annu Rev Biochem 70:677–701.

Kahle J, GP Zauke. 2003. Bioaccumulation of trace metals in the Antarctic amphipod *Orchomene plebs*: evaluation of toxicokinetic models. Mar Environ Res 55:359–384.

Klerks P. 2002. Adaptation, ecological impacts, and risk assessment: Insights from research at Foundry Cove, Bayou Trepagnier, and Pass Fourchon. Human Ecol Risk Assess 8:971–982.

Lee B-G, Griscom SB, Lee J-S, Choi H-J, Koh C-H, Luoma SN, Fisher NS. 1995. Influence of dietary uptake and reactive sulfides on metal bioavailability from aquatic sediments. Science 287:282–284.

Luoma SN, Rainbow PS. 2005. Why is metal bioaccumulation so variable? Biodynamics as a unifying concept. Environ Sci Technol 39:1921–1931.

MacLean RS, Borgmann U, Dixon DG. 1996. Bioaccumulation kinetics and toxicity of lead in *Hyalella azteca* (Crustacea, Amphipoda). Can J Fish Aquat Sci 53:2212–2220.

McDonald DG, Wood CM. 1993. Branchial mechanisms of acclimation to metals in freshwater fish. In: Rankin JC, Jensen FB, editors. Fish ecophysiology. London, UK: Chapman and Hall, p. 297–321.

McGeer JC, Szebedinszky C, McDonald DG, Wood CM. 2000a. Effects of chronic sublethal exposure to waterborne Cu, Cd, or Zn in rainbow trout. 1: Iono-regulatory disturbance and metabolic costs. Aquat Toxicol 50:231–245.

McGeer JC, Szebedinszky C, McDonald DG, Wood CM. 2000b. Effects of chronic sublethal exposure to waterborne Cu, Cd, or Zn in rainbow trout. 2: tissue-specific metal accumulation. Aquat Toxicol 50:245–256.

McGeer JC, Brix KV, Skeaff JM, DeForest DK, Brigham SI, Adams WJ, Green A. 2003. Inverse relationship between bioconcentration factor and exposure concentration for metals: implications for hazard assessment of metals in the aquatic environment. Environ Toxicol Chem 22:1017–1037.

McKim JM. 1994. Physiological and biochemical mechanisms that regulate the accumulation and toxicity of environmental chemicals in fish. In: Hamelink JL, Landrum PF, Bergman HL, Bensen WH, editors. Bioavailability: physical, chemical and biological interactions. Boca Raton, FL: CRC Press, p. 179–201.

Meyer JS, editor. 2005. Diet-borne metal toxicity to aquatic organisms. Pensacola, FL: SETAC Press (in press).

Nott JA, Nicolaidou A. 1990. Transfer of metal detoxification along marine food chains. J Mar Biol Assoc UK 70:905–912.

Nott JA, Nicolaidou A. 1993. Bioreduction of zinc and manganese along a molluscan food chain. Comp Biochem Physiol 104A:235–238.

OECD (Organization for Economic Cooperation and Development). 1996. Guidelines for Testing of chemicals no. 305. Bioconcentration: flow-through fish test. Paris, France.

OECD. 2001. Harmonized integrated hazard classification system for human health and environmental hazards of chemical substances and mixtures. Annex 2. Guidance document on the use of the harmonized system for the classification of chemicals which are hazardous for the aquatic environment. OECD Environment, Health and Safety Publications Series on Testing and Assessment No. 27, ENV/JM/MONO(2001)8, Paris, France.

Paquin PR, Farley K, Santore RC, Kavvadas CD, Mooney KG, Winfield RP, Wu K-B, Di Toro DM. 2003. Metals in aquatic systems: a review of exposure, bioaccumulation, and toxicity models. Pensacola, FL: SETAC Press.

Philips DJH, Rainbow PS. 1988. Barnacles and mussels as biomonitors of trace elements: a comparative study. Mar Ecol Progr Ser 49:83–93.

Phillips DJH, Rainbow PS. 1993. Biomonitoring of trace aquatic contaminants. New York: Elsevier Applied Science, 371 p.

Playle RC, Wood CM 1989. Water chemistry changes in the gill microenvironment of rainbow trout — experimental observations and theory. J Comp Physiol 159B:527–537.

Rainbow PS. 2002. Trace metal concentrations in aquatic invertebrates: why and so what? Environ Pollut 120:497–507.

Rainbow PS, Blackmore G. 2001. Barnacles as biomonitors of trace metal availabilities in Hong Kong coastal waters: changes in space and time. Mar Environ Res 51:441–463.

Reinfelder JR, Wang W-X, Luoma SN, Fisher NS. 1997. Assimilation efficiencies and turnover rates of trace elements in marine bivalves: a comparison of oysters, clams and mussels. Mar Biol 129:443–452.

Reinfelder JR, Fisher NS, Luoma SN, Nichols JW, Wang W-X. 1998. Trace element trophic transfer in aquatic organisms: a critique of the kinetic model approach. Sci Tot Environ 219:117–135.

Schlekat CE, Decho AW, Chandler GT. 1999. Dietary assimilation of cadmium associated with bacterial exopolymer sediment coatings by the estuarine amphipod Leptocheirus plumulosus: effects of Cd concentration and salinity. Mar Ecol Progr Ser 183:205–216.

Schlekat CE, Decho AW, Chandler GT. 2000. Bioavailability of particle-associated silver, cadmium, and zinc to the estuarine amphipod Leptocheirus plumulosus through dietary ingestion. Limnol Oceanogr 45:11–21.

Schlekat CE, Lee B-G, Luoma SN. 2001. Dietary metals exposure and toxicity to aquatic organisms: Implications for ecological risk assessment. In: Newman MC, Roberts MH, Hale RC, editors. Coastal and estuarine risk assessment. Boca Raton, FL: CRC Press, p. 151–188.

Schlekat CE, B-G Lee, SN Luoma. 2002. Assimilation of selenium from phytoplankton by three benthic invertebrates: effect of phytoplankton species. Mar Ecol Progr Ser 237:79–85.

Simkiss K, Taylor MG. 1995. Transport of metals across membranes. In: Tessier A, Turner A, editors. Metal speciation and bioavailability in aquatic systems. Chichester, UK: John Wiley and Sons, p. 2–44.

Skorupa JP, Ohlendorf HM. 1991. Contaminants in drainage water and avian risk thresholds. In: Dinar A, Zilbeman D, editors. The economics and management of water and drainage in agriculture. Boston, MA: Kluwer, p. 345–368.

Turner A. 1996. Trace-metal partitioning in estuaries: importance of salinity and particle concentration. Mar Chem 54:27–39.

USEPA (U.S. Environmental Protection Agency). 1999a. Persistent bioaccumulative toxic (PBT) chemicals; lowering of reporting thresholds for certain PBT chemicals; addition of certain PBT chemicals; community right-to-know toxic chemical reporting. Fed Reg 64:58665–58753.

USEPA. 1999b. Screening level ecological risk assessment protocol for hazardous waste combustion facilities. Appendix E: toxicity reference values. EPA 530-D-99-001A. Washington, D.C.: Office of Solid Waste.

USEPA. 2002a. SAB review of the metals action plan. EPA-SAB-EC-LTR-03-001. Washington, D.C.: Science Advisory Board.

USEPA 2002b. National Recommended Water Quality Criteria: 2002. EPA-822-R-02-047. Washington, D.C.

Van Ginneken L, Chowdhury MJ, Blust R. 1999. Bioavailability of cadmium and zinc to the common carp, *Cyprinus carpio*, in complexing environments: a test for the validity of the free ion activity model. Environ Toxicol Chem 18:2295–2304.

Verbost PM, Senden MH, van Os CH. 1987. Nanomolar concentrations of Cd^{2+} inhibit Ca^{2+} transport systems in plasma membranes and intracellular Ca^{2+} stores in intestinal epithelium. Biochim Biophys Acta 902:247–252.

Vercauteren K, Blust R. 1996. Bioavailability of dissolved zinc to the common mussel *Mytilus edulis* in complexing environments. Mar Ecol Progr Ser 137:123–132.

Wallace WG, Lopez GR, Levinton JS. 1998. Cadmium resistance in an oligochaete and its effect on cadmium trophic transfer to an omnivorous shrimp. Mar Ecol Progress Ser 172:225–237.

Wang, W-X. 2002. Interactions of trace metals and different marine food chains. Mar Ecol Progr Ser 45:46–52.

Wang W-X, Fisher NS. 1996. Assimilation of trace elements and carbon by the mussel *Mytilus edulis*: effects of food composition. Limnol Oceanogr 41:197–207.

Wang W-X, Fisher NS. 1999. Delineating metal accumulation pathways for marine invertebrates. Sci Tot Environ 237/238:459–472.

Wang W-X, Fisher NS, Luoma SN. 1996. Kinetic determinations of trace element bioaccumulation in the mussel *Mytilus edulis*. Mar Ecol Progr Ser 140:91–113.

White SL, Rainbow PS. 1985. On the metabolic requirements for copper and zinc in mollusks and crustaceans. Mar Environ Res 16:215–229.

Williams RJP, da Silva JJRF. 2000. The distribution of elements in cells. Coordinat Chem Rev 200-202:247–348.

Wilson RH, De Eds F, Cox AJ Jr. 1941. Effects of continued cadmium feeding. J Pharmacol Exp Ther 71:222–235.

Wilson RW, Wilson JM, Grosell M. 2002. Intestinal bicarbonate secretion by marine teleost fish — why and how? Biochim Biophys Acta 1566:182–193.

Wood CM. 2001. Toxic responses of the gill. In: Schlenk DW, Benson WH, editors. Target organ toxicity in marine and freshwater teleosts, volume 1 — organs. Washington, D.C.: Taylor & Francis, p. 1–89.

5 Aquatic Toxicity for Hazard Identification of Metals and Inorganic Metal Substances

Andrew S. Green, Peter M. Chapman,
Herbert E. Allen, Peter G.C. Campbell,
Rick D. Cardwell, Karel De Schamphelaere,
Katrien M. Delbeke, David R. Mount,
and William A. Stubblefield

5.1 INTRODUCTION

This chapter deals with toxicity, specifically, harmful effects arising from exposure of biota to metals and inorganic metal substances (collectively referred to as metals). The focus of this chapter is the aquatic environment; it considers exposure from the water column, from sediment, and from ingestion of food or sediment. Exposure of terrestrial wildlife is considered separately in Chapter 6.

To allow incorporation of toxicity into risk-based ranking, prioritization, and screening assessments (referred to as categorization), there must be a means of aggregating toxicological data into a form that effectively expresses the toxicological potency of metals. The aggregation of metals' toxicity data must be sensitive to issues affecting their quality, applicability, and interpretation. There are many factors that affect metal toxicity, the most important being chemical speciation and bioavailability. In addition to these 2 key factors, the following considerations apply:

- In many regulatory assessments, there is great focus on the most sensitive organisms or end points in an effort to preclude environmental risks. For categorization rather than risk assessment, the approach should not strictly be as conservative as possible but rather as comparable as possible, because the goal is to rank relative hazard or risk across different substances including metals.
- Though metals occur in many forms, their toxicity is expected to relate to a very few dissolved chemical species, primarily the free metal ion.

Evaluation of metal toxicity data is, therefore, centered on characterizing (1) dissolution or transformation yielding dissolved chemical species, and (2) the toxicity of these species, rather than (3) the toxicity of the original metal substance.

- There is no doubt that characteristics such as solubility and transformation (and their kinetics), which are discussed in Chapter 3, will greatly influence the ecological effects that may occur from release of a metal into the environment. These effects are large (orders of magnitude). Failing to consider these issues in categorizing metals will result in significant errors.
- Toxicological data vary in quality and reliability. For metals where ample data are available, quality of individual test results should be considered, and data of poor quality should be excluded. In cases where few data are available, lower quality data may have to be used. Whenever possible, data should be normalized to standard exposure conditions to achieve a data set of comparable values.

To meet the data needs of the unit world model (UWM) outlined in Chapter 3, the toxicity data analysis must define benchmark concentrations in various environmental media that correspond to a specified level of biological effect for the specific pathways by which organisms may be exposed. This chapter has 3 main objectives: (1) addressing critical issues related to the appropriate use of toxicity data for categorization, (2) providing input to the UWM, and (3) providing an interim solution to the use of aquatic toxicity data in metal categorization, independent of and in advance of the UWM.

5.2 DATA ACCEPTABILITY

The goal of characterization is often to evaluate and compare the relative hazard/risk of different compounds, whether inorganic or organic, not to derive safe concentrations. Regardless of whether existing or newly generated data are used, all data should be normalized to a standard set of tests conditions, for example, bioavailability or common hardness (Meyer 1999). The ultimate objective is to assess the toxicity of the metal species rather than that of the original metal substance.

5.2.1 DATA EVALUATION AND SPECIES SELECTION CRITERIA

Toxicity data of the highest quality must be used in categorization based on both relevance and reliability. Data relevance relates to the intended use of the data, and whether the test design was appropriate for that use. Data reliability is related to the test methods and the conditions under which the test was conducted, the quality assurance procedures used, whether clear exposure–response relationships were observed, and how well test results were reported. Uncensored and nonscreened toxicity data from the literature should not be used (Batley et al. 1999). Standardized (national and international) experimental designs and methodologies (protocols) should be used to promote comparability of test results.

TABLE 5.1

Examples of Interpretative Consequences to Various Combinations of Data-Poor and Data-Rich Toxicity Results for Metal Compounds

Data Quantity	Interpretation
No data available	Material assumed, worst-case, to be highly toxic
1 acute/chronic value for one or more organisms	Use lowest value available
2 or more acute/chronic values for same organism	Use lowest geometric mean value available (e.g., genus mean value)
10 or more acute/chronic values (for different organisms)	Use species sensitivity distribution (SSD) or effect measure distribution (EMD) approach

Note: The use of acute or chronic values will be determined based on the specific, applicable regulatory framework. However, potentially an acute to chronic factor could be applied to available acute data, allowing for comparison with chronic data.

For categorization, the overall goal is to ensure substance comparability. Therefore, comparable measurement end points should be used for metal toxicity tests. As long as the same end points and metrics are used, it should be possible to reach conclusions regarding relative hazard/risk among materials. The measurement end points should reflect biological relevance on a population basis and not be subjective in nature. Traditionally, this has been interpreted as end points relating to the survival, growth, and reproduction of an organism. Statistical metrics must also be comparable. LC_{50} values are favored for acute tests and EC_x (rather than NOEC, no-observed-effect-concentration) values for chronic test end points.

Studies that are recognized to have substantial (fatal) shortcomings must be rejected even if they provide the lowest reported effect level. When high-quality data are unavailable, and data with shortcomings must be used, these data and the resulting decisions must be clearly identified as uncertain. Procedures must permit the replacement of flawed data with higher-quality data, regardless of whether or not the material is shown to be more or less toxic than originally suggested.

In general, where data are available from chronic toxicity tests, these data should be used preferentially because the mode of action may be different for acute and chronic effects. Comparisons based on chronic toxicity may result in different relative rankings of metals than those based on acute data. However, acute toxicity data are more abundant and are frequently used for categorization because they allow for assessment of a broader range of substances.

Categorizations can be improved by using high-quality data (Table 5.1). Where only 1 or 2 data points exist, and the data are of acceptable quality, it is not unreasonable to use the lowest value in a precautionary manner to derive an environmental no-effect level. However, where a large data set allows a more detailed examination of the potential for adverse effects, all of the data should be used rather than requiring the use of the lowest value. A species sensitivity distribution (SSD) approach is recommended. For this approach, use of 10 or more data points is preferable. Use of 20 data points ensures that, at the fifth percentile level, the number derived is not lower than the lowest value in the data set (Hanson and Solomon 2002; Wheeler et al. 2002). Where multiple valid data points are available for the same end point on the same species, the geometric mean should be calculated and used in the categorization.

Metal substances with large toxicity databases should not be penalized, such as by the use of excessive safety factors. Evaluation systems should reflect greater uncertainty for those materials considered data-poor, and less uncertainty for substances that are data-rich. The results of categorizations based on these 2 types of toxicity information should be labeled accordingly, such as "acceptable" or "interim."

It is recommended for the UWM that environmental effect concentrations be selected in a comparable and consistent manner across metals, without introducing undesirable bias. Use of the UWM will require use of threshold effect concentrations in various media (water, sediment, and soil) to assess potential for effects in each compartment. A key difficulty is the variable quality and quantity of existing metal toxicity data. Use of a consistent approach across metal substances is clearly desirable.

5.2.2 CULTURE AND TEST CONDITIONS

5.2.2.1 Background and Essentiality

Background concentrations of both essential (e.g., Ca, Co, Cu, Fe, and Mg — required by all organisms; B, Mn, Mo, and Ni — required by some organisms; Cd — required by phytoplankton [Lee et al. 1995; Lane et al. 2005]) and nonessential metals (e.g., Hg, Pb) should be measured both prior to and during toxicity testing because these metals have the potential to modify biological responses to toxicants. Deficiencies of essential metals in culture and test water may influence sensitivity to some metals (Caffrey and Keating 1997; Fort et al. 1998; Muyssen and Janssen 2001a, 2001b) (Figure 5.1). Algal culture media often have virtually no bioavailable or free Zn because of the use of EDTA (ethylenediaminetetraacetic acid) in the culture medium (Muyssen and Janssen 2001a), and thus may be Zn-deficient for some algal species.

Preexposure to essential and nonessential metals may trigger increased tolerance as a result of acclimation. Organisms acclimated to low Zn concentrations are more sensitive when exposed to higher Zn concentrations, supporting the link between homeostatic mechanisms (for example, metallothioneins) and metal toxicity/detoxification, which has been demonstrated numerous times (e.g., Depledge and Rainbow 1990). Daphnid EC_{50} values have been shown to vary as a function of different levels of Zn in the culture media (Table 5.2). Existing data suggest that organism metabolic requirements for and homeostasis of Zn are tied to its toxicological sensitivity (Figure 5.1 and Figure 5.2).

Homeostatic responses underlying acclimation include changes in uptake and depuration rates (McGeer et al. 2003), increased production of metallothioneins (Benson and Birge 1985), conversion of metals into inert granules, or a combination of these phenomena (Rainbow 2002). Data suggest the responses are often short-term (days) and reversible (Dixon and Sprague 1981; Muyssen and Janssen 2002), but can be large enough to affect categorization. Cadmium and the essential metals

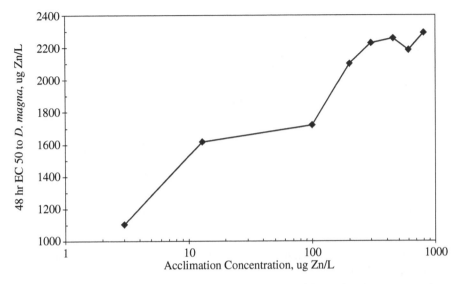

FIGURE 5.1 Toxicity of Zn to *Daphnia magna* as a function of Zn acclimation concentration. (Adapted from Muyssen BTA, Janssen CR. 2001b. Environ Toxicol Chem 20:47-80. With permission.)

TABLE 5.2
Dissolved Zinc Concentrations Measured in Standard Toxicity Test Media Compared to the Average Ambient Background Concentrations of Dissolved Zinc (μg/l)

Source	Type	Dissolved Zn, μg/l
Chu nº10	Algal culture medium	0[a]
Fraquil	Algal culture medium	0.3[a]
ISO and OECD	Test media	1.4[a]
ASTM and EPA	Test media	1.6[a]
World	Ambient	3.25[b]
Northern European lowlands	Ambient	18.5[b]

Source: [a] From Table 2.2 of Muyssen BTA, Janssen CR. 2001a. Chemosphere 45:507–514. [b] Mean values from Zuurdeeg W. et al. 1992. Natuurlijke Achtergrond gehalten van zware metalen en enkele andere sporenelementen in Nederlands oppervlaktewater. Geochem-Research, Utrecht (in Dutch).

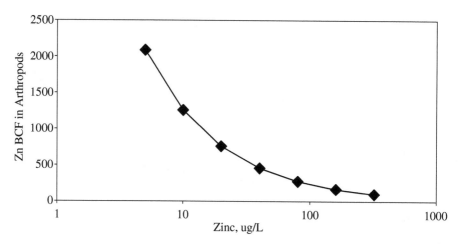

FIGURE 5.2 Relationship between Zn and arthropod BCF. (From Table 3 in McGeer JC. et al. 2003. Environ Toxicol Chem 22:1017–1037. With permission.)

such as Cu and Zn often compete for the same biotic ligands in aquatic organisms (Paquin et al. 2000).

Organisms need species-specific optimal concentration ranges for major ions (e.g., Ca, Mg). For standard test organisms, the ranges of acceptable culture and test conditions (e.g., pH, hardness) as specified within their respective test guidelines should, therefore, be respected. For nonstandard test organisms, species-specific physiological requirements must be reflected in the culture and test conditions. These may have to be defined with further investigation. Purchased or field-collected organisms should be thoroughly acclimated to laboratory water quality because shifts in water quality parameters (e.g., hardness, pH) affect organism fitness and metals toxicity (Meador 1993). Test conditions and culture conditions should be similar. This is often not the case in reported literature.

In summary, the quality of toxicity test data should be checked for validity to see whether: (1) the test organisms have been cultured, collected, or tested in water that is metal deficient, (2) the test water is unrepresentative of natural background for the region under consideration, or (3) sensitive indices of health and performance are compromised relative to organisms held in water of suitable quality. Note that these considerations are of more importance (unless gross differences occur) for detailed ecological risk assessment than for categorization.

5.2.2.2 Other Relevant Test System Characteristics

Abiotic factors controlling metal toxicity should also be within the range of normal field water characteristics, and must be both monitored and controlled. The physicochemical parameters that are considered important for evaluation of the toxicity of metal substances (Ca^{++}, Mg^{++}, H^+, Na^+, CO_3^-, HCO_3^{2-}, SO_4^{2-}, Cl^-) and (oxy)anions (CO_3^{2-}, HCO_3, SO_4^{2-}, Cl^-, OH^-, PO_4^{3-}) are discussed in Section 5.5.

It is recommended that if only one set of water quality characteristics is to be tested for categorization, the physicochemical characteristics of the toxicity test media should correspond to the 50th percentile values of the applicable water quality conditions to avoid extremes. Where appropriate, models (e.g., BLM [biotic ligand model], WHAM [Windemere humic aqueous model]) can be used to estimate effects of free metal ion concentrations in different test media normalized to define test conditions. This allows for the evaluation of alternate water quality characteristics, makes use of a larger portion of the published data, and reduces uncertainties in the toxicity characterization. The ranges of physicochemical characteristics of a large number of European natural waters are described in Table 5.3 and can be useful to define test water characteristics acceptable for categorization. Similar information exists for waters in other geographical areas (the United States) (Erickson 1985).

Special consideration should be given to pH buffering and dissolved organic carbon (DOC) to allow for appropriate interpretation of metal toxicity results. Shifts in physicochemical characteristics during static toxicity testing (e.g., pH drift) that influence metal bioavailability and, hence, data interpretation, can be avoided through buffering (for example, the use of noncomplexing buffers or CO_2 buffering), or flow-through testing (Janssen and Heijerick 2003). DOC is widely recognized to complex metals and alter toxicity results. Ma et al. (1999) demonstrated the influence of metal–DOC complexation kinetics on the toxicity of copper and showed that an equilibration time of 24 hours between metal addition and organism exposure in a toxicity test would be appropriate for natural waters or DOC-containing artificial test media. Note that, if toxicity results are expressed in terms of the free metal ion, the result will be applicable in both DOC-free and DOC-containing media. This approach assumes the free metal ion is responsible for the toxicity; however, if DOC affects metal toxicity by mechanisms in addition to metal complexation (Campbell et al. 1997), then this approach has limitations.

5.2.2.3 Algal Tests

For metals, strong metal-chelating agents should be avoided in toxicity test media (Janssen and Heijerick 2003). EDTA, a strong metal-chelating agent, is a standard constituent of the OECD (Organization for Economic Cooperation and Development) algal test medium used to avoid Fe precipitation and deficiency. Addition of an environmentally relevant amount of naturally less-complexing DOC to algal tests has been considered. Heijerick et al. (2002a) reported that control algal growth was not affected when EDTA was replaced with Aldrich humic acids having the same carbon concentration as EDTA, but the generality of this result is yet to be demonstrated. Modifying the EDTA/Fe ratio or expressing the results as free metal ions are other possible alternatives.

5.3 SEDIMENT EFFECT THRESHOLDS

Because many metals released into the environment will be deposited in aquatic sediments, exposure to contaminated sediment is an important consideration in evaluating potential metal hazards. Existing worldwide guidelines for assessments of the potential toxicity of sediment-associated metals comprise 2 general types: empirically and mechanistically derived values (Batley et al. 2005).

TABLE 5.3
Environmental Distributions of Physicochemical Parameters in European Rivers (1991 to 1996) Data, from the Global Environmental Monitoring System (GEMS)/Water Database (http://www.gemswater.org/publications/index-e.html)

Cumulative Distribution	pH	DOC (mg/l)	Ca (mg/l)	Mg (mg/l)	Na (mg/l)	K (mg/l)	Cl (mg/l)	SO$_4$ (mg/l)	Alkalinity (mg/l CaCO$_3$)
	Nonparametric	LogLogistic	Beta	Gamma	Lognorm	Gamma	Lognorm	Lognorm	Beta
5th Percentile	6.9	2.09	8.10	1.53	3.26	0.13	2.18	6.89	2.98
10th Percentile	7	2.36	13.39	2.14	4.70	0.30	3.90	10.16	5.57
50th Percentile	7.8	4.09	51.20	5.74	17.15	2.44	30.45	39.84	82.05
90th Percentile	8.1	9.27	103.4	12.13	62.57	8.88	237.7	156.29	305.5
95th percentile	8.2	12.79	115.5	14.52	90.31	11.73	425.6	230.3	362.0

Source: From Heijerick DG. et al. 2003. ZEH-WA-02, Report prepared for the International Lead Zinc Research Organization (ILZRO), 34 p. With permission.

Empirically derived guidelines are generally developed from large databases of paired sediment chemistry and toxicity data from field-collected sediments containing complex mixtures of contaminants (Ingersoll et al. 2001). Data are arrayed according to increasing chemical concentration, and then guideline values are selected based on the distribution of effect (toxic) and no-effect data relative to chemical concentration (e.g., the 50th percentile of toxic samples). Using this approach, sediment quality guidelines (SQGs) have been developed for a number of sediment contaminants, including several metals (Ingersoll et al. 2001). Although empirically derived SQGs are capable of segregating sediments into groups with differing probabilities of toxicity, they do not intrinsically reflect causal relationships between specific metals and sediment toxicity and, as a result, are not useful for categorizing metal sediment toxicity.

The second type of SQGs that are mechanistically derived, may have more utility in metals categorization. Mechanistic SQGs developed to date are based on equilibrium partitioning (EqP) theory (van der Kooij et al. 1991; Ankley et al. 1996; Di Toro et al. 2001; USEPA 2002). The basic tenet of EqP theory is that the toxic potency of sediment-associated chemicals is proportional to their chemical activity, which in turn is proportional to their concentration in the sediment. At equilibrium (steady state), interstitial water measurements may be used to estimate chemical activity and have been shown to predict toxicity. The EqP approach has been evaluated in a large number of sediment tests (Berry et al. 1996; Hansen et al. 1996) and has been effective in categorizing sediments as to the likelihood that one of several specific metals (Cu, Cd, Zn, Pb, Ni, and Ag) will cause toxicity in sediments. Metals were shown to not cause toxicity to benthic organisms when concentrations of metals in interstitial water were below effect thresholds determined from water-column toxicity tests. In developing SQG for bulk sediments, safe metal concentrations in sediment have been calculated either on the basis of acid-volatile sulfide (AVS) precipitation with metals (Di Toro et al. 1992, Ankley et al. 1996) or use of whole sediment K_D values to predict interstitial water concentrations (van der Kooij et al. 1991).

For the UWM, application of the EqP approach for sediment categorization can be done by comparing water-column toxicity benchmarks to the concentration of metal present in interstitial water, as predicted from fate calculations. The BLM can be used to predict organic-carbon-normalized metal bioavailability in interstitial water (Di Toro et al. 2005). The use of combined toxicity data for water column and benthic organisms to predict effects on benthic organisms is supported by a lack of statistical differences in the sensitivity of pelagic and benthic/epibenthic organisms when evaluated for a number of different environmental contaminants (USEPA 2002). It should be noted that the EqP approach applies only to divalent metals and silver and does not account for bioaccumulation. Further, the chemical fate of (oxy)anionic metals in sediments is poorly understood. It is likely that different sediment characteristics (other than AVS and OC) determine the overall availability of these metals.

5.4   DIETARY EXPOSURE

Hazards to aquatic organisms historically have been assessed on the basis of toxicity tests conducted using water exposure to metals. However, accumulation of metals by aquatic organisms can occur via both dietary and water exposures (Griscom et al. 2000, 2002; Hare et al. 2003; Meyer et al. 2005; Chapter 4, this volume).

Although combined uptake of metals from water and dietary exposures may contribute to whole-body burden in an approximately additive manner (Luoma 1989; Luoma and Fisher 1997; Barata et al. 2002 — but see Szebedinszky et al. 2001; Kamunde et al. 2002), there are clear examples where metal tissue residues associated with toxicity from water exposure are much lower than those showing no effect when based on dietary exposure (compare Mount et al. 1994 and Marr et al. 1996), as well as the reverse (Hook and Fisher 2001). Such differences are probably attributable to differences in sorption at the gill and kinetics of uptake and internal distribution of metal accumulated via the diet. In any event, they illustrate the difficulties in establishing robust residue–effect relationships across exposure routes and organisms.

Presently, for categorization, bioaccumulation predictions and critical body residues should be used for those metals where they are understood (organoselenium and methylmercury). For those metals where the consequences of dietary exposure are not as well understood (i.e., Cu, Zn, Cd, Ni, and Pb), categorization for aquatic organisms should continue to be based on assessment of water exposure only, with incorporation of dietary exposure and critical residue concepts as advancing science allows. Note, however, there have been no demonstrations of effects in the field from dietary exposure to metals other than organoselenium and methylmercury except in cases where there were historical exceedances of national water quality criteria/guidelines. Thus, there is no clear evidence that categorization of other metals without considerations of dietary exposure will lead to egregious error.

5.5 BIOAVAILABILITY

There is extensive evidence that total metal concentrations are poor predictors of metal bioavailability or toxicity in water (Campbell 1995; Bergman and Dorward-King 1997; Janssen et al. 2000; Paquin et al. 2002; Niyogi and Wood 2004), soil (Chapter 6), and sediment (Ankley et al. 1996). The first key step in evaluating inorganic metal bioavailability is to recognize the importance of metal speciation, both physically (dissolved vs. particulate metal) and chemically (free metal ions vs. complexed metal forms), as some metal forms (species) intrinsically have different toxicological potencies.

5.5.1 SPECIATION

Metal speciation has been determined to be an important factor in determining bioavailability and uptake/toxicity to aquatic organisms. Additionally, the computation of metal partitioning among dissolved and particulate forms (e.g., using the Surface Chemistry Assemblage Model for Particles (SCAMP) — Lofts and Tipping 1998, 2000, 2003), and within the dissolved phase among the free metal ion, inorganic and organic complexes is important. In each case, a crucial question to be addressed in evaluating toxicity is how to relate solution inorganic chemistry and chemical activities of various metal forms (that is, the metal speciation) to metal uptake and toxicity. Current approaches utilize the WHAM (Tipping 1998; http://windermere.ceh.ac.uk/aquatic%5Fprocesses/wham/whamtitlebar.htm) as a current state-of-the-science speciation model that predicts the extent of binding between dissolved metals and natural organic matter. It has been calibrated for a large number of cationic metals over a wide range of environmental conditions, and has been adopted as the speciation component of the BLM. The current use of WHAM 5 (Tipping 1994) in the BLM construct, however, does not preclude the future use of other types of speciation models, such as WHAM 6 (introduced in 2002) or nonideal competitive adsorption (NICA) (Kinniburgh et al. 1996).

5.5.2 Biotic Ligand Model (BLM)

The BLM has been gaining increased interest in the scientific and regulatory community for predicting and evaluating metal bioavailability and toxicity due to its ability to account for both metal speciation in the exposure medium (through WHAM) and competition between toxic metal species and other cations (Ca^{2+}, Mg^{2+}, Na^{2+}, and H^+) at the organism–water interface. This concept was originally developed for fish species (Di Toro et al. 2001) by combining knowledge on metal speciation (Tipping 1994), metal binding (and competition) on fish gills (Playle et al. 1992, 1993), and the relation between gill-bound metal and toxicity (MacRae et al. 1999). Concurrent with model development, research has focused on elucidating the BLM's physiological processes and mechanistic underpinnings (Grosell et al. 2002). The BLM construct for gill-breathing organisms assumes that metal ions bind to ion transporters and disturb ion balances within the organism.

Inspired by these early efforts, BLMs have been developed that can predict the acute toxicity of a number of cationic metals to a large number of freshwater (gill-breathing) organisms (Table 5.4). In addition to advances in acute toxicity assessment, the BLM approach has been demonstrated to reduce bioavailability-related uncertainty of chronic toxicity threshold values for an important number of biota (Delbeke and Van Sprang 2003). Additional research is being done in this important area.

5.5.3 Algae

The mechanisms forming the basis of the BLM-framework for gill-breathing organisms (that is, disturbance of ion-balance) cannot necessarily be extrapolated to algal species. The interaction of a metal with an algal cell will normally involve the following steps: (1) diffusion of the metal from the bulk solution to the biological surface, (2) sorption/surface complexation of the metal at passive binding sites within the protective layer, or at sites on the outer surface of the plasma membrane, and (3) uptake or internalization of the metal (transport across the plasma membrane). The incoming metal encounters a wide range of potential binding sites, which can usefully be divided into 2 classes: *physiologically inert* sites, where the metal may bind without obviously perturbing normal cell function, and *physiologically active* sites, where the metal affects cell metabolism. In the latter case, metal binding may affect cell metabolism directly, for example, if the binding site corresponds to a

TABLE 5.4
Some Available Aquatic Bioavailability Models

Metal	Species	Reference	Remark
Cu	*Pimephales promelas*	Santore et al. (2001)	
	Daphnia magna	De Schamphelaere et al. (2002)	Acute, also other BLM calibrated to limited data set by Santore et al. (2002)
		De Schamphelaere et al. (2003); De Schamphelaere and Janssen (2004a)	Chronic
	Daphnia pulex	Santore et al. (2001)	
	Ceriodaphnia dubia	Santore et al. (2001)	
	Pseudokirchneriella subcapitata	De Schamphelaere et al. (2003)	Chronic (72 h)
Zn	*Oncorhynchus mykiss*	Santore et al. (2002)	Acute
		De Schamphelaere and Janssen (2004b)	Chronic
	Pimephales promelas	Santore et al. (2001)	
	Daphnia magna	Heijerick et al. (2002a)	Acute, also BLM calibrated to limited data set by Santore et al. (2002)
		Heijerik et al. (2005)	Chronic
	Pseudokirchneriella subcapitata	Heijerick et al. (2002b); De Schamphelaere et al. (2005)	
Cd	*Oncorhynchus mykiss*	Santore et al. (2002)	
	Pimephales promelas	Santore et al. (2002)	
Ni	*Pimephales promelas*	Wu et al. (2003)	
	Oncorhynchus mykiss	Wu et al. (2003)	
	Daphnia magna	Wu et al. (2003)	
	Ceriodaphnia dubia	Wu et al. (2003)	
Pb	*Oncorhynchus mykiss*	MacDonald et al. (2002)	MINEQL+ as speciation model
Ag	*Oncorhynchus mykiss*	Paquin et al. (1999)	
	Daphnia magna	Bury et al. (2002)	
	Daphnia pulex	Bury et al. (2002)	

Note: Unless noted otherwise, models predict acute metal toxicity; Nigoyi and Wood (2004) provide a more comprehensive summary and discussion of existing BLMs.

membrane-bound enzyme, or indirectly, if the bound metal is subsequently transported across the plasma membrane into the cell. Once within the cell, the metal may interact with a variety of intracellular sites, resulting in positive or negative consequences (Campbell 1995; Campbell et al. 2002).

Within the BLM construct, the physiologically active sites at the cell surface constitute the algal biotic ligand. Empirical bioavailability models have been developed and validated for the green alga *Pseudokirchneriella subcapitata* (also known as *Selenastrum capricornutum* and *Raphidocelis subcapitata*) to predict toxicity of

Cu (De Schamphelaere et al. 2003) and Zn (Heijerick et al. 2002a) under a wide range of environmental conditions. Although competing cations (Ca^{2+}, Mg^{2+}, Na^+) may play a significant role, the most important determinants of algal toxicity of these two metals are pH and DOC.

5.5.4 BLM Data Gaps and Future Directions

Less effort has been spent on attempting to understand the bioavailability of (oxy)anionic metal ions (for example, molybdate, selenate, vanadate, arsenate, and chromate). A predictive (BLM-type) approach has not been developed. Bioavailability modeling is required for (oxy)anionic metals to complement the extensive knowledge base developed for cationic metals. Based on common chemical logic, the bioavailability of (oxy)anionic metal ions will probably be determined by different water quality characteristics than for cationic metals. For example, competition may come from anions such as phosphate (for arsenate uptake — Wang et al. 2002) or sulfate (for selenate uptake) (Terry et al. 2000). Complexation by organic matter will not be important, but pH (protonation/deprotonation equilibria) will also affect speciation.

5.5.5 Taking Bioavailability into Account

As a first approximation, the relative solubility of a metal substance in water indicates its relative hazard for categorization purposes. For substances that simply dissolve, either yielding the intact parent compound or dissociating into component ions, their equilibrium aqueous solubility will be a useful guide. If the substance undergoes transformation (for example, the case for metal sulfides or oxides), then the rate of dissolution/transformation becomes important (Chapter 3).

Once the metal is in solution, the worst-case or default scenario would be that it remains entirely free, or uncomplexed. For many metal cations, this assumption would significantly overestimate bioavailability. In a typical receiving water, the cation would form inorganic and organic complexes (M–Cl, M–CO_3, M–SO_4, M–DOM) with a consequent decrease in bioavailability. The tendency of cations to form such complexes varies markedly from metal to metal and, to a lesser extent, will vary from one receiving medium to another. Equilibrium modeling (MINEQL, WHAM) can be used to take these speciation differences into account. A possible application would be to normalize the data from the toxicity testing literature, expressing the various toxicology end points (e.g., LC_{50}, EC_{50}) not in terms of total dissolved metal, but rather as the free metal cation concentration in the test medium (Batley et al. 2002). However, this approach would not take into account the protective effects exerted by the hardness cations or the H^+ ion, nor would it account for metal species other than the free metal (i.e., Cu^{++} vs. $CuOH^+$). The BLM allows such refinements.

In some situations, and for some metals, assuming the free metal ion is responsible for toxicity, the toxicity may be over- or underestimated, as exemplified in Figure 5.3. On the left panel of the figure a (mechanistic) relation between $EC_xMe_y^+$ and pH is given. This relation represents the commonly accepted competition

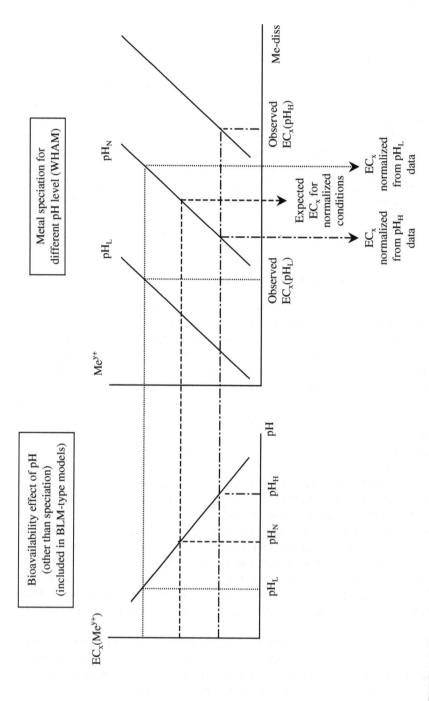

FIGURE 5.3 Example of normalization of EC_x values to a given condition.

between cationic metals and protons for biological interfaces, resulting in lower $EC_xMe_y^+$ at higher pH levels. In fact, a correct normalization of toxicity data is only possible when these competition effects are taken into account (for example, through the BLM). However, when only correcting toxicity data for complexation (that is, normalizing to free metal ion activity), under- or overestimation of expected toxicity values may arise. The same inferences may be made for other cations related to, for example, hardness. Toxicity is underestimated when corrected to a lower cation concentration and *vice versa*. The effects of cation competition may vary widely across metals and across biota. Thus, this uncertainty in categorizing metals for which no BLM-type models are available needs to be resolved.

In comparison with metal cations, the bioavailability of metal (oxy)anions (AsO_4, CrO_4, MoO_4, SeO_4, and VO_4) is less well understood (Section 5.5.4). Given our current lack of knowledge, (oxy)anions, once in solution, should conservatively be considered 100% bioavailable unless data are available to suggest otherwise.

5.6 INTEGRATED APPROACH FOR RISK/HAZARD ASSESSMENTS USING TOXICITY

Most of the toxicity data available for metals have been generated using soluble metal salts (e.g., $CdCl_2$) because of the ease of getting the metal substance being tested into solution. Current categorization frameworks normally use the toxicity data from soluble metals salts to characterize the toxicity of all metal compounds (e.g., Cd-metal, $CdCO_3$). This assumes that all metal elements and compounds will ultimately transform and solubilize from their initial forms into free metal ions at the same level (and rate) as the corresponding soluble metal salts, which leads to inaccuracies as most metal-containing substances are sparingly soluble (Allen and Batley 1997). Consequently, application of toxicity data from soluble metal salts to categorize sparingly soluble metals is inappropriate (Adams et al. 2000). To facilitate metal comparisons and ensure discrimination between metals, a risk-based categorization approach was developed to link toxicity data for soluble metal salts to their respective metals. This approach can be integrated into the UWM described in Chapter 3, and could provide an intermediate step for categorization of metal-containing substances based on toxicity, until the UWM is validated and accepted.

5.6.1 APPROACH

The following stepwise approach enables development of a metals categorization index based on a toxicity–solubility relationship (cf. Figure 5.4):

1. A toxicity value is identified for a metal in question (ideally, based on the soluble, free metal ion; realistically, based on dissolved metal concentrations). This, in general, is M^{n+} or MO_y^{z-}. The toxicity value should be an LC_{50}, EC_{50}, or other effect-based value, expressed as the free (or dissolved) metal (cf. Section 5.2 and Section 5.5). Because toxicity values are also dependent on solution composition (e.g., pH), they should be corrected

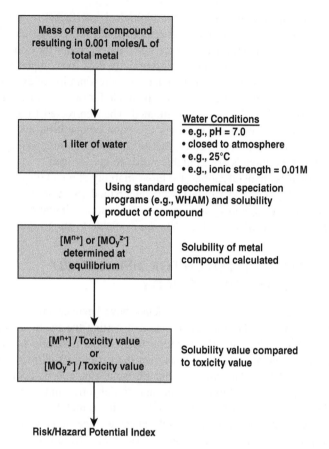

FIGURE 5.4 Schematic showing development of a metals categorization index based on toxicity–solubility (cf. Section 5.6.1, this volume).

for the water quality conditions used in the assessment. The pH should be the same as used for the solubility calculation.

2. A mass is defined for the compound of interest that would result in 0.001 mol/l of total metal. A fixed number of moles rather than a fixed mass for all compounds facilitates comparison of potential risk for different compounds of the same metal and for comparison among metals.

3. This mass is introduced into 1 liter of water possessing, for example, fixed pH (7.0) and ionic strength (0.01 M). Note, more than 1 water type could be used. The water is closed to the atmosphere and is at a specified temperature (e.g., 25°C). A mass of 0.001 mol is sufficient to provide excess solid compound for sparingly soluble compounds.

4. Using the solubility product of the compound, $[M^{n+}]$ or $[MO_y^{z-}]$ is calculated at equilibrium. For many metal compounds, the calculation can be performed using a standard geochemical speciation program such as MINTEQA2 (Allison et al. 1991) or MINEQL+ (Schecher and McAvoy

1992). The computation is performed without permitting any other compounds to precipitate. Solubility products and other equilibrium constants not in the program's default thermodynamic database should be taken from other databases (e.g., NIST [U.S. National Institute of Science and Technology]) or from published literature. If no reliable data can be found, the solubility of the compound should be determined in the laboratory for specified conditions. Standard conditions should be used, and care must be taken to ensure that any remaining solids have been completely separated from the solution prior to analysis.

MINEQL+ and MINTEQA2 are perhaps the most common computer programs used to solve speciation problems, particularly those involving precipitation and solubilization reactions. However, WHAM VI is better at computing the complexation of metals with natural organic matter (Tipping 1998). The major difference among computer programs is in the quality of the thermodynamic databases used. Errors in the databases have been found (Serkiz et al. 1996). Thus, the databases should be reviewed to ensure data quality. Further, the programs should be run by an individual with a good understanding of chemistry to ensure that the results are reasonable and realistic.

Although metal elements tend to have low solubility, they may corrode, giving rise to corrosion products that have finite solubility. The soluble metal concentration arising from corrosion processes cannot be calculated with confidence. Because the extent and rate of corrosion are highly dependent on physical (e.g., particle size, surface imperfections, flow) and chemical conditions (e.g., pH, oxidants, DOC), corrosion should be determined for environmentally relevant conditions. Measurement of the soluble metal after an appropriate reaction period (e.g., 7 d) should be used for calculation of free metal ion present (or alternatively soluble metal).

5. Compare $[M^{n+}]$ or $[MO_y^{z-}]$/"toxicity value" as an index to categorize metals.

5.6.2 EXAMPLES

The approach outlined in Section 5.6.1 was applied to various metal compounds (Table 5.5). Solubility (moles dissolved M/l) was calculated with MINEQL 4.5 software for arbitrary fixed conditions: $[M]_T = 0.01\ M$; pH = 7; ionic strength = 0.01 M. The metal and the appropriate anion (e.g., Cl^-, CO_3^{2-}, S^{2-}) were introduced as components. Dissolved solids (Type V species as defined by the software) were inspected and a single form was chosen (e.g., $CdCl_2$, $ZnCO_3$). The model was then run and from the output tables, two values were extracted: the total dissolved metal (Table 5.5), and the free metal ion concentrations, $[M^{z+}]$. Note that a value of 0.01 M in column 2 corresponds to 100% solubility. In most cases $[M^{z+}] \approx [M]_T$, but in some cases, the calculated free-ion concentration was much less than the total dissolved metal. This situation may arise either because the anion that enters the solution phase with the metal subsequently forms soluble complexes (e.g., in the cases of $CdCl_2$ and $ZnSO_4$), or because the metal itself forms polynuclear complexes

TABLE 5.5
Toxicity Categorization Index Example Output

Metal	Solubility (mol/l M)	Solubility (mol/l $Mz+$)	Toxicity Value (μg/l)	Toxicity Value (mol/l M)	Hazard Potential Index
$CdCl_2$	0.01	0.005	0.25	2.224E09	4496400
$Cd(OH)_2$	0.01	0.0081			4496400
$CdCO_3$	0.0000677	0.0000657			30441
$CuCl_2$	0.01	0.0006	9	1.416E07	70607
$CuCO_3$	0.000175	0.000088			1236
$PbCl_2$	0.01	0.00198	3.2	1.544E08	647500
$Pb(OH)_2$	0.00000254	0.00000214			164
$NiCl_2$	0.01	0.00967	52	8.860E07	11287
$NiCO_3$	0.01	0.00664			11287
$ZnCl_2$	0.01	0.00957	120	1.836E06	5447
$ZnCO_3$	0.000673	0.000655			367
$AgNO_3$	0.01	0.01	0.12	1.102E09	9071833
$AgCl$	0.0000152	0.0000148			13789
Ag_2S	0.00000685	8.72E22			6214
$HgCl_2$	0.01	0.000315	0.91	4.53662E09	2204286
$Hg(NO_3)_2$	0.000229	0.000229			50478
HgS	3.63E09	2.07E38			3.63E07

(e.g., in the cases of $CuCl_2$ and $PbCl_2$). Note, however, that the reference toxicity value (column 4 and column 5; EPA Water Quality Criteria chosen for the current example) is expressed in terms of total dissolved metal (not the free metal cation), and thus the categorization index is currently calculated as the quotient [dissolved metal]/[toxicity value] (column 2/column 5). Values for the free metal ion concentration in column 3 are thus given for information purposes only.

5.7 CONCLUSIONS AND RECOMMENDATIONS

Based on the basic principles associated with toxicity testing and data interpretation for metals, there is clear need for an integrative approach to evaluate metal hazard for categorization. The UWM is described as such an approach in Chapter 3. Three principles are set forth to ensure that robust and reliable toxicity data are applied in the UWM in relevant environmental compartments. First, test conditions should be normalized (e.g., similar temperatures) and described. Second, the same measurement end points should be used (ideally survival, growth, and fecundity which reflect population-level effects). Third, toxicity should be reported in terms of comparable metrics (e.g., EC_x values).

Developing methods for inorganic metal hazard assessment and comparative ranking requires the following:

- Data should be screened for quality before use in categorization. Data recognized as having fatal shortcomings should be rejected outright. Other data should be categorized as "acceptable" or "interim," depending on their quality. Similar qualifications apply to categorizations based on those data.
- The lowest available toxicity value should not be used when an integrative approach is possible. Standardized approaches that normalize data sets based on data quality should be used.
- The water quality from which the test organisms were captured, cultured, and tested should be defined and be similar to the test medium, with no deficiencies or excesses of essential metals.
- For categorization of metals in sediments, pore water concentrations can be used in conjunction with aquatic toxicity values derived from tests of water column and benthic organisms.
- Bioavailability should be used to normalize data sets, reducing uncertainty and increasing comparability.
- Until the UWM is validated, categorization of metals based on toxicity should rely on integration of toxicity and solubility data, based ideally upon free metal ion concentrations or, less ideally, upon dissolved metal concentrations.
- Dietary uptake can be a major source of metal body burden for some metals. However, the bioreactivity of inorganic metals within aquatic organisms remains poorly understood, and there is presently no clear evidence that water quality guidelines are not protective for both water and dietary exposures to inorganic metals.

ACKNOWLEDGMENT

We acknowledge verbal contributions from Amy Crook (Center for Science in Public Participation in Affiliation with Environmental Mining Council, Victoria, B.C., Canada) during the initial workgroup meetings in Pensacola, FL.

REFERENCES

Adams WJ, Conard B, Ethier G, Brix KV, Paquin PR, Di Toro DM. 2000. The challenges of hazard identification and classification of insoluble metals and metal substances for the aquatic environment. Human Ecol Risk Assess 6:1019–1038.

Allen HE, Batley GE. 1997. Kinetics and equilibria of metal-containing materials: ramifications for aquatic toxicity testing for classification of sparingly soluble metals, inorganic metal compounds and minerals. Human Ecol Risk Assess 3:397–413.

Allison JD, Brown DS, Novo-Gradac KJ. 1991. MINTEQA2/PRODEFA2, a geochemical assessment model for environmental systems: Version 3.0 user's manual. EPA/600/3-91/021.

Ankley GT, Di Toro DM, Hansen D, Berry WJ. 1996. Technical basis and proposal for deriving sediment quality criteria for metals. Environ Toxicol Chem 15:2056–2066.

Barata C, Markich SJ, Baird DJ. 2002. The relative importance of water and food as cadmium sources to *Daphnia magna* Straus. Aquat Toxicol 61:143–154.

Batley GE, Apte SC, Stauber JL. 1999. Acceptability of aquatic toxicity data for the derivation of water quality guidelines for metals. Mar Freshwater Res 50:729–738.

Batley GE, Burton GA Jr, Chapman PM, Forbes VE. 2002. Uncertainties in sediment quality weight-of-evidence (WOE) assessments. Human Ecol Risk Assess 8:1517–1547.

Batley GE, Stahl Jr. RG, Babut MP, Bott TL, Clark JR, Field LJ, Ho KT, Mount DR, Swartz RC, Tessier A. 2005. The scientific underpinnings of sediment quality guidelines. In: Wenning RJ, Batley GE, Ingersoll, CG, Moore DW, editors. Use of sediment quality guidelines (SQGs) and related tools for the assessment of contaminated sediments. Pensacola, FL: SETAC Press, p. 39–119.

Benson WH, Birge WJ. 1985. Heavy metal tolerance and metallothionein induction in fathead minnows: results from field and laboratory investigations. Environ Toxicol Chem 4:209–217.

Bergman HL, Dorward-King EJ. 1997. Reassessment of metals criteria for aquatic life protection. Pensacola, FL: SETAC Press.

Berry WJ, Hansen DH, Mahony JD, Robson DL, Di Toro DM, Shipley BP, Rogers B, Corbin JM, Boothman WS. 1996. Predicting the toxicity of metal-spiked laboratory sediments using acid-volatile sulfide and interstitial water normalizations. Environ Toxicol Chem 15:2067–2079.

Bury NR, Shaw J, Glover C, Hogstrand C. 2002. Derivation of a toxicity-based model to predict how water chemistry influences silver toxicity to invertebrates. Comp Biochem Physiol C 133:259–270.

Caffrey PB, Keating KI. 1997. Results of zinc deprivation in daphnid culture. Environ Toxicol Chem 16:572–575.

Campbell PGC. 1995. Interactions between trace metals and aquatic organisms: a critique of the free-ion activity model. In: Tessier A, Turner DR, editors. Metal speciation and bioavailability in aquatic systems. Chichester, UK: John Wiley & Sons, p. 45–102.

Campbell PGC, Twiss MR, Wilkinson KJ. 1997. Accumulation of natural organic matter on the surfaces of living cells: implications for the interaction of toxic solutes with aquatic biota. Can J Fish Aquat Sci 54:2543–2554.

Campbell PGC, Errécalde O, Fortin C, Hiriart-Baer VP, Vigneault B. 2002. Metal bioavailability to phytoplankton — applicability of the Biotic Ligand Model. Compar Biochem Physiol 133:189–206.

Depledge MH, Rainbow PS. 1990. Models of regulation and accumulation of trace metals in marine invertebrates. Comp Biochem Physiol 97C:1–7.

De Schamphelaere KAC, Janssen CR. 2004a. Development and field validation of a biotic ligand model predicting chronic copper toxicity to *Daphnia magna*. Environ Toxicol Chem 23:1365–1375.

De Schamphelaere KAC, Janssen CR. 2004b. Bioavailability and chronic toxicity of zinc to juvenile rainbow trout (*Oncorhynchus mykiss*): comparison with other fish species and development of a biotic ligand model. Environ Sci Technol 38:6201–6209.

De Schamphelaere KAC, Heijerick DG, Janssen CR. 2002. Refinement and field validation of a biotic ligand model predicting acute copper toxicity to *Daphnia magna*. Comp Biochem Physiol C 133:243–258.

De Schamphelaere KAC, Vasconcelos FM, Heijerick DG, Tack FMG, Delbeke K, Allen HE, Janssen CR. 2003. Development and field validation of a predictive copper toxicity model for the green alga *Pseudokirchneriella subcapitata*. Environ Toxicol Chem 22:2454–2465.

De Schamphelaere KAC, Lofts S, Janssen CR. 2005. Bioavailability models for predicting acute and chronic toxicity of zinc to algae, daphnids and fish in natural surface waters. Environ Toxicol Chem 24:1190–1197.

Delbeke K, Van Sprang P. 2003. Integration of environmental bioavailability of metals in risk assessment — the copper example. Abstract Book, 13th Annual SETAC Europe Meeting, Hamburg, Germany.

Di Toro DM, Mahony JD, Hansen DJ, Scott KJ, Carlson AR, Ankley GT. 1992. Acid volatile sulfide predicts the acute toxicity of cadmium and nickel in sediments. Environ Sci Technol 26:96–101.

Di Toro DM, Allen HE, Bergman H, Meyer JS, Paquin PR, Santore CS. 2001. Biotic ligand model of the acute toxicity of metals. 1. Technical basis. Environ Toxicol Chem 20:2383–2396.

Di Toro DM, McGrath JA, Hansen DJ, Berry WJ, Paquin PR, Mathew R. Wu KB, Santore RC. 2006. Predicting sediment metal toxicity using a sediment Biotic Ligand Model: methodology and initial application. Chemistry 24, in press.

Dixon, DG, Sprague JB. 1981. Acclimation to copper by rainbow trout *(Salmo gairdneri)*— a modifying factor in toxicity. Can J Fish Aquat Sci 38:880-888.

Erickson RJ. 1985. Analysis of major ionic content of selected US waters and application to experimental design for the evaluation of the effect of water chemistry on the toxicity of copper. Duluth, MN: Environmental Research Laboratory, Office of Research and Development, U.S. Environmental Protection Agency.

Fort DJ, Propst TL, Schetter T, Stover EL. 1998. Teratogenic effects of insufficient boron, copper and zinc in *Xenopus*. Teratology 57:252.

Griscom SB, Fisher NS, Luoma SN. 2000. Geochemical influences on assimilation of sediment-bound metals in clams and mussels. Environ Sci Technol 34:91–99.

Griscom SB, Fisher NS, Luoma SN. 2002. Kinetic modeling of Ag, Cd and Cu bioaccumulation in the clam *Macoma balthica*: quantifying dietary and dissolved sources. Mar Ecol Progr Ser 240:127–141.

Grosell M, Nielsen C, Bianchini A. 2002. Sodium turnover rate determines sensitivity to acute copper and silver exposure in freshwater animals. Comp Biochem Physiol C 133:287–303.

Hansen DJ, Berry WJ, Mahony JD, Boothman WS, Di Toro DM, Robson DL, Ankley GT, Ma D, Yan Q, Pesch CE. 1996. Predicting the toxicity of metal contaminated field sediments using interstitial water concentrations of metals and acid volatile sulfide normalizations. Environ Toxicol Chem 15:2080–2094.

Hanson ML, Solomon KR. 2002. New technique for estimating thresholds of toxicity in ecological risk assessment. Environ Sci Technol 36:3257–3264.

Hare L, Tessier A, Borgmann U. 2003. Metal sources for freshwater invertebrates: pertinence for risk assessment. Human Ecol Risk Assess 9:779–793.

Heijerick DG, De Schamphelaere KAC, Janssen CR. 2002a. Biotic ligand model development predicting Zn toxicity to the alga *Pseudokirchneriella subcapitata*: possibilities and limitations. Comp Biochem Physiol C Toxicol Pharmacol 133:207–218.

Heijerick DG, De Schamphelaere KAC, Janssen CR. 2002b. Predicting acute zinc toxicity for *Daphnia magna* as a function of key water chemistry characteristics: development and validation of a biotic ligand model. Environ Toxicol Chem 21:1309–1315.

Heijerick DG, De Schamphelaere KAC, Janssen CR. 2003. Application of biotic ligand models for predicting zinc toxicity in European surface waters. ZEH-WA-02, Report prepared for the International Lead Zinc Research Organization (ILZRO), 34 p.

Heijerick DG, De Schamphelaere KAC, Van Sprang PA, Janssen CR. 2005. Development of a chronic zinc biotic ligand model for *Daphnia magna*. Ecotox Environ Saf, in press.

Hook SE, Fisher NS. 2001. Sublethal effects of silver in zooplankton: importance of exposure pathways and implications for toxicity testing. Environ Toxicol Chem 20:568–574.

Ingersoll CG, MacDonald DD, Wang N, Crane JL, Field LJ, Haverland PS, Kemble NE, Lindskoog RA, Severn C, Smorong DE. 2001. Predictions of sediment toxicity using consensus-based freshwater sediment quality guidelines. Arch Environ Contam Toxicol 41:8–21.

Janssen CR, Heijerick DG. 2003. Algal toxicity testing for environmental risk assessments of metals: physiological and ecological considerations. Rev Environ Contam Toxicol 178:23-52.

Janssen CR, De Schamphelaere K, Heijerick D, Muyssen B, Lock K, Bossuyt B, Vangheluwe M, Van Sprang P. 2000. Uncertainties in the environmental risk assessment of metals. Human Ecol Risk Assess 6:1003–1018.

Kamunde C, Grosell M, Higgs D, Wood CM. 2002. Copper metabolism in actively growing rainbow trout (*Oncorhynchus mykiss*): interactions between dietary and waterborne copper uptake. J Exp Biol 205:279–290.

Kinniburgh DG, Milne CJ, Benedetti MF, Pinheiro JP, Filius J, Koopal LK, Van Riemsdijk WH. 1996. Metal ion binding by humic acid: application of the NICA-Donnan model. Environ Sci Technol 30:1687–1698.

Lane TW, Saito MA, George GN, Pickering IJ, Prince RC, Morel FMM. 2005. A cadmium enzyme from a marine diatom. Nature 435:42.

Lee JG, Roberts SB, Morel FMM. Cadmium, a nutrient for the marine diatom *Thalassiosira weissflogii*. Limnol Oceanogr 40:1056-1063.

Lofts S, Tipping E. 1995. An assemblage model for cationic binding by natural particulate matter. Geochim Cosmochim Acta 62:2609–2625.

Lofts S, Tipping E. 2000. Solid-solution metal partitioning in the Humber rivers: application of WHAM and SCAMP. Sci Tot Environ 251/252:381–399.

Lofts S, Tipping E. 2003. SCAMP—surface chemistry assemblage model for particles. Centre for Ecology and Hydrology, CEH Windermere, Ferry House, Far Sawrey, Ambleside, Cumbria, UK. (http://www.ife.ac.uk/aquatic_processes/scamp.htm).

Luoma SN. 1989. Can we determine the biological availability of sediment-bound trace elements? Hydrobiologia 176/177:379–396.

Luoma SN, Fisher N. 1997. Uncertainties in assessing contaminant exposure from sediments. In: Ingersoll CG, Dillon T, Biddinger G, editors. Ecological risk assessments of contaminated sediments. Pensacola, FL: SETAC Press, p. 211–237.

Ma H, Kim SD, Cha DK, Allen HE. 1999. Effects of kinetics of complexation by humic acid on the toxicity of copper to *Ceriodaphnia dubia*. Environ Toxicol Chem 18:828–837.

MacDonald A, Silk L, Schwartz M, Playle RC. 2002. A lead–gill binding model to predict acute lead toxicity to rainbow trout (*Oncorhynchus mykiss*). Comp Biochem Physiol C 133:227–242.

MacRae RK, Smith DE, Swoboda-Colberg N, Meyer JS, Bergman HL. 1999. The copper binding affinity of rainbow trout (*Oncorynchus mykiss*) and brook trout (*Salvelinus fontinalis*) gills: implications for assessing bioavailable metal. Environ Toxicol Chem 18:1180–1189.

Marr JCA, Lipton J, Cacela D, Hansen JA, Bergman HA, Meyer JS, Hogstrand C. 1996. Relationship between copper exposure duration, tissue copper concentration, and rainbow trout growth. Aquat Toxicol 36:17–30.

McGeer JC, Brix KV, Skeaff JM, Deforest DK, Brigham SI, Adams WJ, Green A. 2003. Inverse relationship between bioconcentration factor and exposure concentration for metals: implications for hazard assessment of metals in the aquatic environment. Environ Toxicol Chem 22:1017–1037.

Meador JP. 1993. The effect of laboratory holding on the toxicity response of marine infaunal amphipods to cadmium and tributyltin. J Exp Mar Biol Ecol 174:227–242.

Meyer JS. 1999. A mechanistic explanation for the $\ln(LC_{50})$ vs. $\ln$(hardness) adjustment equation for metals. Environ Sci Technol 33:908–912.

Meyer JS, Adams WJ, Brix KV, Luoma SN, Mount DR, Stubblefield WA, Wood CM, editors. 2005. Toxicity of diet-borne metals to aquatic biota. Pensacola, FL: SETAC Press, in press.

Mount DR, Barth AK, Garrison TD, Barten KA, Hockett JR. 1994. Dietary and waterborne exposure of rainbow trout (*Oncorhynchus mykiss*) to copper, cadmium, lead, and zinc using a live diet. Environ Toxicol Chem 13:2031–2041.

Muyssen BTA, Janssen CR. 2001a. Zinc acclimation and its effect on the zinc tolerance of *Pseudokirchneriella subcapitata* and *Chlorella vulgaris*. Chemosphere 45:507–514.

Muyssen BTA, Janssen CR. 2001b. Multigeneration zinc acclimation and tolerance in *Daphnia magna*: implications for water quality guidelines and ecological risk assessment. Environ Toxicol Chem 20:47–80.

Muyssen BTA, Janssen CR. 2002. Accumulation and regulation of zinc in *Daphnia magna*: links with homeostasis and toxicity. Arch Environ Contam Toxicol 43:492–496.

Niyogi S, Wood CM. 2004. Biotic ligand model, a flexible tool for developing site-specific water quality guidelines for metals. Environ Sci Technol 38:6177–6192.

Paquin PR, Di Toro DM, Santore RC, Trivedi D, Wu KB. 1999. A biotic ligand model of the acute toxicity of metals. III. Application to fish and *Daphnia* exposure to silver. In: Integrated approach to assessing the bioavailability and toxicity of metals in surface waters and sediments. USEPA-822-E-99-001. Washington, D.C.: U.S. EPA Science Advisory Board, Office of Water, Office of Research and Development, p. 3–59, 3–102.

Paquin PR, Santore RC, Wu KB, Kavvadas CD, Di Toro DM. 2000. The biotic ligand model: a model of the acute toxicity of metals to aquatic life. Environ Sci Policy 3:S175–S182.

Paquin PR, Gorsuch JW, Apte S, Batley GE, Bowles KC, Campbell PGC, Delos CG, Di Toro DM, Dwyer RI, Galvez F, and others. 2002. The biotic ligand model: a historical overview. Comp Biochem Physiol C 133:3–36.

Playle RC, Gensemer RW, Dixon DG. 1992. Copper accumulation on gills of fathead minnows: influence of water hardness, complexation and pH of the gill microenvironment. Environ Toxicol Chem 11:381–391.

Playle RC, Dixon DG, Burnison K. 1993. Copper and cadmium binding to fish gills: estimates of metal-gill stability constants and modelling of metal accumulation. Can J Fish Aquat Sci 50:2678–2687.

Rainbow PS. 2002. Trace metal concentrations in aquatic invertebrates: why and so what? Environ Pollut 120:497-507.

Santore RC, Di Toro DM, Paquin PR, Allen HE, Meyer JS. 2001. Biotic ligand model of the acute toxicity of metals. 2. Application to acute copper toxicity in freshwater fish and *Daphnia*. Environ Toxicol Chem 20:2397–2402.

Santore RC, Mathew R, Paquin PR, Di Toro D. 2002. Application of the biotic ligand model to predicting zinc toxicity to rainbow trout, fathead minnow and *Daphnia magna*. Comp Biochem Physiol C 133:271–285.

Schecher WD, McAvoy DC. 1992. MINEQL+: a software environment for chemical equilibrium modeling. Comput Environ Urban Syst 16:65–76.

Serkiz SM, Allison JD, Perdue EM, Allen HE, Brown DS. 1996. Correcting errors in the thermodynamic database for the equilibrium speciation model MINTEQA2. Water Res 30:1930–1933.

Szebedinszky C, McGeer JC, McDonald DG, Wood CM. 2001. Effects of chronic Cd exposure via the diet or water on internal organ-specific distribution and subsequent gill Cd uptake kinetics in juvenile rainbow trout (*Oncorhynchus mykiss*). Environ Toxicol Chem 20:597–607.

Terry N, Zayed AM, de Souza MP, Tarun AS. 2000. Selenium in higher plants. Ann Rev Plant Phys 51:401–432.

Tipping E. 1994. WHAM — a chemical equilibrium model and computer code for waters, sediments, and soils incorporating a discrete site/electrostatic model of ion-binding by humic substances. Comput Geosci 20:973–1023.

Tipping E. 1998. Humic ion-binding model VI: an improved description of the interactions of protons and metal ions with humic substances. Aquat Geochem 4:3–48.

U.S. Environmental Protection Agency (USEPA). 2002. Equilibrium partitioning sediment guidelines (ESGs) for the protection of benthic organisms: metal mixtures (cadmium, copper, lead, nickel, silver and zinc). EPA-822-R-02-045. Washington, D.C.: Office of Water.

van der Kooij LA, van de Meent D, van Leeuwen CJ, Bruggeman WA. 1991. Deriving quality criteria for water and sediment from the results of aquatic toxicity tests and product standards: application of the equilibrium partitioning method. Water Res 6:697–705.

Wang J, Zhao FJ, Meharg AA, Raab A, Felderman G, McGrath SP. 2002. Mechanisms of arsenic hyper accumulation in *Pteris vittata*: uptake kinetics, interactions with phosphate and arsenic speciation. Plant Physiol 130:1552–1561.

Wheeler JR, Grist EPM, Leung KMY, Morritt D, Crane M. 2002. Species sensitivity distributions: data and model choice. Mar Pollut Bull 45:192–202.

Wu KB, Paquin PR, Navab V, Mathew R, Santore RC, Di Toro DM. 2003. Development of a biotic ligand model for nickel: Phase I. Water Environment Research Foundation, Report 01-ECO-10T, p. 1–37.

Zuurdeeg W, van Enk RJ, Vriend SP. 1992. Natuurlijke Achtergrond gehalten van zware metalen en enkele andere sporenelementen in Nederlands oppervlaktewater. Geochem-Research, Utrecht (in Dutch).

6 Hazard Assessment of Inorganic Metals and Metal Substances in Terrestrial Systems

Erik Smolders, Steve McGrath,
Anne Fairbrother, Beverley A. Hale,
Enzo Lombi, Michael McLaughlin,
Michiel Rutgers, and Leana Van der Vliet

6.1 FOREWORD

The primary focus of this SETAC Pellston Workshop was the aquatic environment. Although the terrestrial environment received consideration with regard to the unit world model (UWM) (Chapter 3) and is discussed in this chapter, this discussion is not intended to provide a comprehensive analysis of the current state of the art for hazard assessment of metals in terrestrial systems.

6.2 INTRODUCTION

Soils are important sinks for metals in the environment. The major routes of metal input to soils are atmospheric deposition, application of animal manures and inorganic fertilizers, and in localized areas, mining and smelting activities, addition of sewage sludge, and alluvial deposition. Over the short term, metals generally have a greater level of adverse effects on biota in aquatic systems than in terrestrial systems because, in terrestrial systems, metals are rapidly bound to soil solids particles (Chapman et al. 2003; Section 6.3.3, this volume). Because metals are less mobile in soils than in aquatic systems, adverse effects in terrestrial systems may only be observed after much longer periods of exposure. Effectively, this means that any hazard assessment scheme for metals should include terrestrial systems when long-term ecosystem sustainability is considered.

6.3 PERSISTENCE OF METALS IN SOIL

6.3.1 RESIDENCE TIME OF METALS IN SOIL

Metals persist in soil due to their high affinity for soil solid phases (Allen 2002). Critical factors affecting the mass balance of metals in soils are the anthropogenic and natural inputs and the outputs via leaching to groundwater and removal through surface erosion and crop harvesting. The elimination half-life of metals in soil $(t_{1/2})$* can be predicted from a soil mass as:

$$t_{1/2} = \frac{0.69 \times d \times 10000}{y \times TF + \dfrac{R}{\rho Kd}}$$

where d is the soil depth in meters, y is the annual crop yield (t ha^{-1} y^{-1}), TF is the ratio of the metal concentration in plants to the concentration in soil, R is net drainage of water out of the soil (m^3 ha^{-1} y^{-1}), ρ is the bulk density of the soil (ton$_{dw}$ m^{-3} or kg$_{dw}$ L^{-1}), and K_D is the ratio of the metal concentration in soil to that in soil solution (L kg$^{-1}_{dw}$). Continuing aerial and other emissions of metals to soils increase soil metal concentrations, such that the time required to achieve 95% of steady state is about 4 half-lives. Selenium (Se), a metalloid usually present in anionic forms in soil, approaches steady state after only 1 year and, as a consequence, Se soil concentrations after 100 years and at steady state are identical (Table 6.1). In contrast, Cu, Cd, Pb, and Cr III do not approach steady state after even 100 years. The soil concentrations of these metals are very similar after 100-years' loading if inputs are identical; however, the steady state concentrations are very different, because the time necessary to approach steady-state is a function of the K_D (at a constant loading rate). Note that the time needed to approach steady state for all the metals in Table 6.1, except Se, is on the order of thousands of years, and it is difficult to envisage that soil conditions would not change in this time frame. This result implies that the concept of a steady state for metals in soil, even if attractive from a conceptual point of view, is elusive.

6.3.2 CRITICAL LOADS OF METALS

A critical load concept can be developed by predicting metal loading rates required to achieve toxic thresholds in soil. Figure 6.1 shows the results of such critical loads (note that they are all reported relative to Cd) after defining maximum permitted soil concentrations (critical loadings) for five metals. Two scenarios are compared: steady state and the 100-year time horizon. In both cases, the critical Se loadings rates are largest, even though its toxic threshold in soil is lowest. This result is related to the large mobility of Se when applied as soluble selenate, that is, the critical load can be large because losses by leaching are large. Loading rates are smallest for Cd due to its high intrinsic toxicity and relatively high K_D. The ranking of critical loads of metals

---

* Time required to reduce the initial concentration by 50% if metal input is zero.

TABLE 6.1
Time to Achieve 95% of Steady-State Metal Concentration in Soil and Total Soil Metal Concentrations after 100 Years and at Steady State

Metal	Loading Rate (g ha^{-1} y^{-1})	K_D (l kg^{-1})	T (years)[a]	Soil Metal Concentration (mg added metal kg^{-1})	
				After 100 Years	Steady State
Se	100	0.3	1.3	0.01	0.01
Cu	100	480	1860	2.4	16
Cd	100	690	2670	2.4	23
Pb	100	19000	73300	2.6	633
Cr III	100	16700	64400	2.6	556

Note: Based on a soil depth of 25 cm, a rain infiltration rate of 3000 m^3 ha^{-1} y^{-1}, and the assumption that background was 0 at the start of loading.

[a] Time to achieve 95% of steady-state metal concentration in soil.

Source: K_D values from De Groot AC. et al. 1998. National Institute of Public Health and the Environment, The Netherlands. Report nr 607220 001. (http://www.rivm.nl/bibliotheek/rapporten/607220001.html), p. 260.

is different when based on either steady-state situations or a fixed timeframe. As an example, the critical load of Pb is about 3 times larger than that of Cu when calculations are made on a 100-year timescale and is due to the larger toxicity threshold for Pb (i.e., it is less toxic). The reverse is true in a steady-state situation, because steady-state metal concentrations (at equivalent input, see Table 6.1) are about 40-fold larger for Pb than for Cu due to the differences in K_D between these 2 metals.

The critical load of Se is much larger than all the other elements considered. However, in this model calculation a large leaching rate, typical of temperate climates, is considered (300 mm y^{-1}). The critical load of Se would be much lower than that of Cu under arid conditions at a fixed time (but not at steady state).

Long-term changes in soil properties — for instance, over 100 years — can drastically affect metal partitioning in soil and thereby metal persistence in this environmental compartment. For example, land use changes resulting in a decrease in soil pH, such as the conversion of arable land to forest, can increase metal mobility by an order of magnitude (Figure 6.2). Thus, hazard-ranking metals in terms of their steady-state critical load is not reliable over long time frames without accounting for potential changes in climate and land use.

6.3.3 AGING OF METALS IN SOIL

Persistence of total metals in soil and persistence of metal bioavailability and solubility are not the same. In the latter case, the process called *aging* is responsible for decreasing metal bioavailability over time (Chapman et al. 2003). Aging is

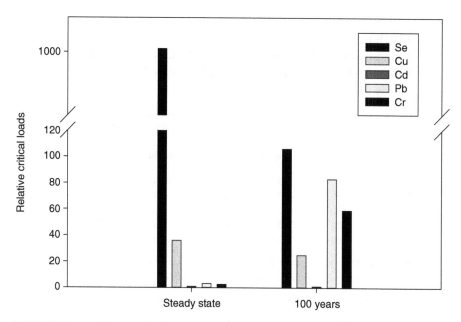

FIGURE 6.1 Relative critical load of metals required to achieve soil ecotoxicological criteria at steady state and after 100 years of metal loading. The Dutch ecotoxicological soil criteria used are (in mg kg^{-1}_{dw}): Se, 0.8; Cu, 40; Cd, 1.6; Pb, 140; Cr III, 100. (From Crommentuijn T. et al. 1997. National Institute of Public Health and the Environment, The Netherlands. Report nr 601501 001. (http://www.rivm.nl/bibliotheek/rapporten/601501001.html), p. 46. With permission.) Other parameters used in the calculation are the same as in Table 6.1. Note that the time to achieve steady state varies by orders of magnitude (see Table 6.1).

defined as the slow reactions that occur following rapid partitioning of added soluble metals between solution and solid phases in soil, which can take years to attain equilibrium. These slow reactions remove metals from the labile pool to a fixed pool. The mechanisms are ascribed to micropore diffusion, occlusion in solid phases by (co)precipitation, isomorphous substitution in crystal lattices, and cavity entrapment. Evidence of aging processes is provided by studies of metal extractability and lability. Easily extractable metal pools, experimentally added to soils in the form of soluble salts, revert with time ($\geq$ 1 year) to more strongly bound forms. For example, Hamon et al. (1998) measured the rate of aging of Cd in agricultural soils where this metal was added as a contaminant in phosphate fertilizers. Using a radioisotopic technique, they developed a model that estimated Cd aging on the order of 1 to 1.5% of the total added Cd per year. Using a similar technique, Young et al. (2005) studied the fixation of Cd and Zn in 23 soils amended with inorganic metal complexes over a period of 811 days. They observed that the extent of aging increased with soil pH (Figure 6.3).

Aging reactions follow reversible first-order kinetics and are clearly dependent upon pH. The proportion of Zn that remains labile after > 1-year aging appears to gravitate to a mean value of approximately 30% for soils with pH > 6.5 (Figure 6.3). There is an apparent reversibility of fixed metal as determined by the isotopic

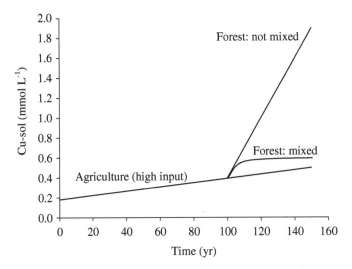

FIGURE 6.2 Predicted changes in soil solution Cu concentrations as a result of land use changes affecting soil organic C levels. Three scenarios are presented (see Moolenaar et al. 1998 for further details): (1) high input agriculture, (2) high input agriculture following by afforestation and litter mixed into the soil, and, (3) high input agriculture followed by afforestation and litter not mixed into the soil.

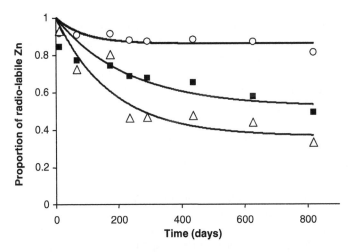

FIGURE 6.3 Time-dependent reduction in radio-lability of Zn in 23 soils incubated for over 800 d. The soils are grouped into 3 pH ranges: 6 soils pH < 5.5 (■); 10 soils pH 5.5 to 6.5 (○); 7 soils pH > 6.5 (△). Solid lines are the fit of a reversible first-order kinetic equation to each grouped dataset. (Reprinted from Young S. et al. 2006. Isotopic dilution methods. In: Hamon RE, McLaughlin MJ, editors. Natural attenuation of trace element availability in soils. Pensacola, FL: SETAC Press. With permission.)

dilution method (Young et al. 2005). Data fitting in Figure 6.3 was performed using a reversible kinetic model, which requires a final equilibrium position with less than 100% fixation of metal. This is supported by the fact that, in field soils, either contaminated or not, there is a substantial degree of metal lability. For instance, Degryse et al. (2003) investigated the lability of Zn in a range of field collected soils and found that labile Zn typically varied between 10 to 40% of the total and was dependent on soil pH. Aging reactions as assessed by isotopic dilution techniques are effectively reversible.

Ma et al. (2005) investigated the aging of Cu in 19 European soils using an isotopic dilution technique. Their results showed that the lability of Cu added to soils rapidly decreased after addition, especially in the soils with pH > 6.0, followed by a slow decrease in Cu lability. The lability of Cu added to soils also decreased with increasing incubation temperature. The soil and environmental factors governing attenuation rates were: soil pH, organic matter content, incubation time, and temperature. The attenuation of Cu lability was modeled on the basis of 3 processes: precipitation/nucleation of Cu on soil surfaces, Cu occlusion within organic matter, and diffusion of Cu into micropores.

Information regarding the relative importance of aging reactions for different metals and metalloids is limited. Aging reactions can affect both partitioning of metals in soil and assessment of critical toxicity values in soil when these are based on total metal concentrations. Increased aging enhances metal retention by the soil-solid phase. Consequently, if partitioning is calculated measuring the total and pore-water concentration of a metal in well-equilibrated soils, an aging factor is already included in the calculation. However, when K_Ds are calculated from adsorption isotherms, aging must be considered separately.

Assessment of threshold toxicity values, including soil quality guidelines, is influenced by aging, because toxicological tests are usually performed during the period of relatively fast metal fixation that follows metal addition to soil. From Figure 6.3, it can be predicted that Zn toxicity based on total soil concentrations derived from tests conducted in high pH soils immediately after addition of inorganic metal salts would be greater than toxicity derived from tests conducted after 1 year.

6.3.4 TRANSFORMATION OF SPARINGLY SOLUBLE COMPOUNDS

Metals often enter soils not in dissolved form but as sparingly soluble compounds. Dissolution of these compounds is related to chemical and physical properties characteristic of both the compounds and the soils. Environmental parameters such as temperature and humidity have a strong influence on any metal transformations. Without knowledge of its dissolution rate, the hazard posed by a specific compound cannot be correctly assessed. Dissolution of sparingly soluble compounds in soil is often different from that observed in water because soils provide a sink for the reaction products of dissolution. The buffering capacity of soil is also greater than that of aquatic systems, as are moisture conditions and oxygen content. Finally, aging reactions in soils may take place at the same time as dissolution. For instance, the amount of soluble vanadium (V) is larger when the source is sodium vanadate than in the case of vanadium oxide (Table 6.2). This differential dissolution is even

TABLE 6.2
Concentration (mg/l) of Co and V in Pore Water of a Sandy Soil Amended with 2 Different Compounds and Incubated for 24 Weeks

	Time of Incubation (weeks)			
Compound	2	4	12	24
Co_3O_4	0.003	0.005	0.002	0.002
$CoCl_2$	9.6	9.3	10.5	11.2
V_2O_5	19.8	13.3	9.9	7.5
Na_3VO_4	40.7	21.4	7.8	12.0

Note: Addition rates were 100 and 250 mg kg^{-1} for Co and V, respectively. Smolders and Degryse (unpublished data).

more pronounced when cobalt oxide and chloride are compared (Table 6.2). However, in the case of V the aging rate is larger than the dissolution rate so that soil pore water concentrations decrease with time, whereas the concentration of soil pore water Co does not decrease over time.

The problem posed by sparingly soluble compounds in soil can be addressed using a toxicological approach that includes some typical processes involved in compound transformation in soil, that is, dissolution, partitioning, aging, and so on (McLaughlin et al. 2002). In this case, 3 parallel toxicity tests were suggested. The first is performed after a short equilibration time (2 to 7 days). The remaining 2 tests are performed after a prolonged equilibration time (60 days) with and without a leaching step after 2 to 7 days. The leaching step is included to remove the toxicity of counterions released during dissolution. If toxicity increases over time, then hazard classification has to take into account transformation rates, and the substance may be reclassified into a more hazardous category. Additional details of this approach, which is recommended for general use, are provided in Section 6.5.

6.4 BIOACCUMULATION OF METALS IN THE TERRESTRIAL FOOD CHAIN

6.4.1 DEFINING BIOACCUMULATION FACTOR (BAF) AND BIOCONCENTRATION FACTOR (BCF) IN THE TERRESTRIAL ENVIRONMENT

For terrestrial ecosystems, bioaccumulation is the basis of two ecologically important outcomes: primary phyto- or zootoxicity and secondary toxicity to animals feeding on contaminated plants and animals. Such measurements typically involve BAFs (bioaccumulation factors) or BCFs (bioconcentration factors). Problems associated with using these measures generically for metals are detailed in Chapter 4. Specific issues related to the terrestrial environment are described below; additional details regarding invertebrates is provided in Allen (2002).

For vegetation, BAF is defined as field measurements of metal concentration in plant tissues divided by metal concentration in soil (or soil solution); BCF is defined as the same measurement carried out in artificial media in the laboratory. These ratios are similarly determined for aquatic organisms; BAF by default includes dietary exposure, whereas BCF does not. The BAFs for plants may include aerially deposited metals to shoots as well as soil particles adhering to roots, depending on the preparation of field samples before analysis, which should not be part of the BCFs determined in hydroponic culture. Although these surface-adhered fractions of the BAF are not likely to be phytotoxic for metals, they will contribute to trophic transfer of metals; their removal from plant tissues before tissue metal analyses is rare, and when it does occur, it is likely to be incomplete, although how incomplete is unclear. For soil invertebrates, similar differences in these ratios apply. The BCFs with earthworms may not include additional feeding of the animals during the study. For higher order organisms (for example, birds and mammals), whole-body BAFs generally are not calculated, with the exception of small mammals (Sample et al. 1998a). Rather, concentrations in target tissues are measured for comparison to toxic levels (Beyer et al. 1996).

6.4.2 MEASURING BAF/BCFs — THE DENOMINATOR

For terrestrial plants, there have been considerable investigations attempting to determine the best measurement of soil metal that will predict metal bioaccumulation. It is beyond doubt that total metal in soil is a poor predictor of metal concentrations in the plant; that is, BAF values expressed based on total metals in soil are highly variable. Several mechanisms have been highlighted as to why this is the case. Plant tissue concentrations of essential metals are maintained within physiological limits over a wide range of total metal concentrations (e.g., Zn, Cu), thereby leading to BAF values that decrease with increasing total metal concentrations. Plant tissue concentrations of nonessential elements depend on the solubility of those elements in soil and on the presence of competing elements in solution. Solution culture studies have shown that the free metal ion is generally absorbed faster than anionic metal complexes, and suggestions have been made that the free ion (activity) in soil solution is a predictor for uptake of metals. Free-ion measurements in soil solution were demonstrated to reduce variability in the BAF for some metals but not others, and for no metal were the BAFs collapsed into one value by using free ion in the denominator (Johnson et al. 2003). Several mechanisms have been proposed to explain why the free ion is not a unique predictor of crop metal concentrations across widely different soils. First, the uptake of a free metal ion is affected by the concentrations of competing ions, that is, H^+, Ca^{2+}, Mg^{2+} and varying concentrations of these ions in solution of different soils obscure the relationship between the free metal ion in solution and that in the crop (Hough et al. 2005). Second, metal uptake increases with increasing concentrations of metal complexes at constant free metal activity, suggesting that either metal complexes can also be taken up by plants or that the complexes overcome diffusive limitations (Smolders et al. 1996; Parker and Pedler 1997; Berkelaar and Hale 2003a, 2003b).

Development of a biotic ligand model (BLM) for plants is improving predictions of metal phytotoxicity by correcting for the competitive inhibition of toxicity by Ca^{2+}, Mg^{2+}, and H^+ in the soil solution (Weng et al. 2003, 2004; Thakali et al. 2005). A BLM for predicting Cd and Zn *uptake* (or BAF) by plants from soil has revealed that protons are the main competing ions for metal uptake (Hough et al. 2005). The BLM does not yet accommodate kinetic limitations to metal uptake, specifically the role of labile metal complexes as buffers of free metal activity at the interface between the biotic ligand and exposure solution which, under conditions of high rates of metal uptake and low free ion activity, can be a zone of depletion.

Uptake of metals from soils by invertebrates is also influenced by metal speciation. However, the relationship is considerably more complex, particularly for hard-bodied species (for example, *Collembola*). Insects and arthropods are exposed primarily through dietary uptake, either through the soil food chain or by direct ingestion of soil particles and soil solution. The relative bioavailability of metals in these 3 compartments contributes to the potential for uptake and storage in the invertebrates. Earthworms and other soft-bodied organisms may also be exposed through dermal uptake as a function of concentrations in soil pore water.

6.4.3 Interpreting BAF/BCFs

Because BAF/BCFs vary with exposure concentration, they cannot be used as a point estimate of hazard, as is common for organic contaminants (Chapter 4). The slopes of the plots (either BAF or BCF vs. exposure concentration or body concentration vs. exposure concentration) can only be used to generalize data, assuming linearity. The slope of the body concentration vs. exposure concentration is a measure of the organism's ability to regulate the metal. Lower slope (less steep) indicates that the organism can better regulate its exposure to the metal, as observed for essential metals (e.g., Zn), whereas steeper slopes are observed for nonessential metals such as Pb (Heikens et al. 2001). The corollary of this is that BAF values show a steeper decline with increasing exposure concentrations for essential than for nonessential metals. The slopes differ by up to an order of magnitude across different orders of invertebrate taxa, and the ranking of taxa in terms of BAF varies among metals (Heikens et al. 2001). Perhaps by improving the resolution of these slopes (for example, further groupings among taxa, including data from plants), common trends among metals could be discerned. However, at present no recommendations are possible regarding interpreting metal BAFs/BCFs in terrestrial systems.

6.4.4 Trophic Transfer Factors

Trophic transfer factors obviously vary because of variable dietary habits of wildlife. These factors, moreover, vary because of variable speciation of metal in the diet. For example, Cd incorporated into leaves is substantially complexed by phytochelatins, but Cd incorporated into seeds of grains is more likely to be complexed in phytates (myoinositol hexaphosphates). Gastric dissolution of phytates is notoriously low, thus metals exposure for animals feeding on foliage might be different than for animals feeding on grains. It is clear that the accumulation of Cd in target organs differs between dietary material into which the Cd was incorporated during growth and the same dietary material to which Cd was added as a soluble salt (Chan et al. 2000, 2004). It is unrealistic to attempt to incorporate this nuance into hazard identification; however, the data in these 2 studies demonstrate that determination of trophic transfer factors by addition of soluble metal salt to diets may lead to overestimations.

6.4.5 Trophic Transfer of Metals

Bioaccumulation of metals on a whole body basis is generally small for wildlife consuming vegetation; those consuming invertebrates may, however, have higher exposures (Sample et al 1998a). Significant sequestration of ingested metals may occur in inert tissues such as bone and hair (Beyer et al. 1996). However, due to dilution and low bioavailability (or ingestion) of metals from inert tissues, there is generally no biomagnification in upper portions of the terrestrial food web.

In the aquatic environment, Hg provides the best example of increased hazard through transformation. However, the environmental conditions necessary for mercury biomethylation in aquatic systems (sulfate-reducing anaerobic bacteria in sediments) exist only to a limited extent in the oxic soils of terrestrial systems. In terrestrial systems, the main Hg issue is not transformation, but intermedia transport, as some plants can act as vectors of Hg^0 transport from the soil to the air (Leonard et al. 1998). Elements such as selenium, tellurium, tin, lead, antimony, bismuth, cadmium, nickel, polonium, thallium, and germanium can potentially methylate under particular environmental conditions (Thayer 2002); however, methylation does not always increase toxicity. Organoarsenicals, for example, are significantly less toxic than their inorganic counterparts (Hindmarsh and McCurdy 1996); therefore, methylation may sometimes be a route to reduce, rather than enhance, hazard.

6.4.6 Proposed Approach for Incorporation of BAF into Hazard Assessment

The key to assessing whether or not movement of a metal through the food web will result in sufficient concentrations to cause problems to wildlife receptors is to compare wildlife dietary thresholds to natural levels of metals in soils and to determine how much the metal would need to increase in the food chain to reach a toxic level. Bioaccumulation factors for metals in plants and invertebrates (the ratio of the concentration in biota to the concentration in soil) can then be compared to the toxicity/soil ratio. If the former is much smaller than the latter, the metal will rank low in regard to potential hazard, whereas if there is only a small difference, then the hazard ranking would be much higher. However, such a comparison is complicated by: (1) variable background concentrations of metal in soils, (2) lack of consensus for derivation of wildlife toxicity threshold values, (3) complexity of dietary estimates, and (4) concentration and soil-type dependence of uptake factor relationships. Therefore, it is suggested that metals be ranked first in terms of relative toxicity to wildlife (Table 6.3) and that the ranking then be modified by the bioaccumulation potential in soil invertebrates and in plants (Table 6.4 and Table 6.5).

TABLE 6.3
Ranking of Dietary Toxicity Thresholds

	Mammal	Bird
Most toxic	Me-Hg	Me-Hg
	Cd	Pb
	Pb	Se
	V	Cr
	Se	Cd
	Hg	As
	As	Hg
	Zn	V
	Cu	Cu
	Ni	Ni
	Cr	Zn
Least toxic	Mn	Mn

Note: There are approximately three orders of magnitude between most and least toxic. Ranked based on dietary toxicity threshold concentrations calculated from toxic dose thresholds. (From USEPA. 1999. Screening level ecological risk assessment protocol for hazardous waste combustion facilities. Available from: www.epa.gov/earth1r6/6pd.rcra_c/protocol/slerap2.htm.) Food intake rates reported by EPA, 1993. (From [USEPA (U.S. Environmental Protection Agency)]. 1993. Exposure factors handbook. Available from: http://cfpub.epa.gov/ncea/cfm/wefh.cfm?ActType=default.)

This approach would produce 2 separate hazard rankings that could be applied to appropriate parts of the terrestrial food web or, alternatively, a single ranking could be developed based on which part of the food web has the highest BAFs (Table 6.6). Such rankings would reflect order of magnitude differences; better discrimination is not currently possible.

6.5 RANKING METAL TOXICITY IN TERRESTRIAL SYSTEMS

Toxicity thresholds of metals in soil, expressed as total concentrations, are well known to vary largely among species, biological endpoints, and the properties of the soil that affect metal bioavailability. Moreover, single species toxicity tests may not be predictive for field conditions, given the complex interactions at the ecosystem level, including acclimation and adaptation processes and differences in metal bioavailability.

Efforts are under way to explain variability of metal toxicity among soils or between metal-spiked and field-contaminated soils, either using the BLM concept

TABLE 6.4
Ranking of Metal Bioaccumulation Potentials in Soil Invertebrates

BAF Slope	Soil Invertebrates[a]	Earthworms Only[b]	Rank
Negative	Cd	Cd	1
Negative	Cu	Cu	2
Zero (constant body concentration)	Zn	Zn	3
Negative	Pb	Pb	4
Negative		Ni	5

Note: 1 = Highest potential.

[a] From Heikens A. et al. 2001. Environ Pollut 113:385–393. With permission.

[b] From Sample BE. et al. 1998b. Development and validation of bioaccumulation models for earthworms. ES/ER/TM-220. Oak Ridge, TN: U.S. Department of Energy, Oak Ridge National Laboratory. With permission.

TABLE 6.5
Ranking of Bioaccumulation Potential of Metals in Plants

BAF Slope	Plants	Rank
Negative	Se	1
Negative	Cd	2
Negative	Zn	3
Zero	Hg	4
Negative	Cu	5
Negative	Pb	6
Negative	As	7
Negative	Ni	8

Note: 1 = Highest potential.

Source: From USDOE (U.S. Department of Energy). 1998. Empirical models for the uptake of inorganic chemicals from soil by plants. BJC/OR-133. Oak Ridge, TN: Oakridge National Laboratory. With permission.

TABLE 6.6

Hazard Ranking of Metals Based on Toxicity Modified by Uptake Factors

Toxicity Ranking						Uptake Ranking				Final Ranking[b]	
Mammal		Bird		Wildlife[a]		Invert		Plant			
Metal	Rank	Metal	Rank	Metal	Rank	Metal	Rank	Metal	Rank	Metal	Rank
Cd	1	Pb	1	Pb	1.5	Cd	1	Se	1	Cd	1.7
Pb	2	Se	2	Cd	2.0	Cu	2	Cd	2	Se	2.0
Se	3	Cd	3	Se	2.5	Zn	3	Zn	3	Pb	3.7
Hg	4	As	4	Hg	4.5	Pb	4	Hg	4	Hg	4.3
As	5	Hg	5	As	4.5	Ni	5	Cu	5	Cu	4.3
Zn	6	Cu	6	Cu	6.5			Pb	6	As	4.3
Cu	7	Ni	7	Zn	7.0			As	7	Zn	5.8
Ni	8	Zn	8	Ni	7.5			Ni	8	Ni	7.0

[a] Rank determined by averaging the ranks of birds and mammals.
[b] Rank determined by reassigning the wildlife ranks 1 to 7 and then averaging ranks of wildlife, plants, and invertebrates.

(Section 6.4.2) or using a regression based approach (Oorts et al. 2006). These empirical models allow normalization of toxicity thresholds for soil properties affecting toxicity and facilitate the ranking of toxicity thresholds of different metals based on existing data. A better method for ranking relative hazard is, however, to perform identical toxicity tests under standard conditions. This method is still not ideal as it depends on generic toxicity tests, usually conducted in the laboratory with standard test organisms. Laboratory toxicity is, as previously discussed (Section 6.3.3), typically higher than toxicity in the real environment for reasons including aging of metals in soils and acclimation, adaptation, and community tolerance (Posthuma et al. 2001; Chapman et al. 2003; Smolders et al. 2004). Toxicity of metals in the environment is best assessed using soil mesocosms or actual field data. Furthermore, effects of metals in the environment are best assessed using a weight of evidence approach that combines the individual lines of evidence of chemical measurements, toxicity determinations, and observations of field communities to decrease uncertainties (Chapman et al. 2002; Fairbrother 2003; Rutgers and Den Besten 2005).

Hazard assessment presently, and for the foreseeable future, primarily depends on soil toxicity tests conducted in the laboratory. This being the case, it is important that such testing include 3 specific trophic levels as a minimum: microbes, invertebrates, and plants. A single test is considered insufficient for hazard assessment because different trophic levels may react differently to substances. These 3 trophic levels represent primary producers, consumers, and decomposers, which are some key elements of the soil ecosystem (Figure 6.4). The microbial test represents the basis of the soil ecosystem in terms of biomass for fueling the food chain; microbes are involved in almost every nutrient cycle. The invertebrate test represents the consumer part (detritivores, carnivores, fungivores, herbivores, etc.) and higher

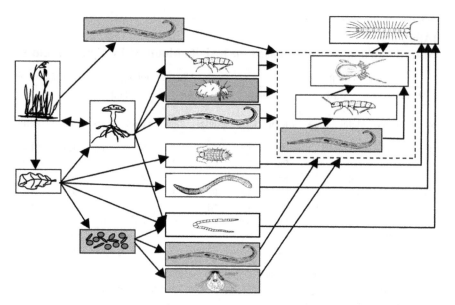

FIGURE 6.4 A soil food web with functional compartments. (Redrawn from De Ruiter, P.C. et al. 1995. *Science* 269:1257–1260. With permission.)

trophic levels in the soil food web. The primary producer (plant) test represents the major input of carbon into the system.

These 3 tests should be run with Cd as the positive (toxic) control so that rankings can be calculated relative to Cd (Figure 6.5). Further, at least 2 soils should be selected for the tests, one that accentuates the bioavailability of cationic metals (pH 5 to 5.5) and the other that maximizes the bioavailability of anionic forms (pH 7.5 to 8). Thus, bioavailability will be taken into account in the hazard testing, albeit nonexplicitly. The output generated will be conservative because it is a reasonable worst-case for the 2 forms of ions. The proposed tests also allow for transformations of insoluble compounds by requiring that they be performed: (1) 2 to 7 days after mixing the test substance into the soil, (2) 60 days after the initial 2 to 7 days incubation, and (3) after leaching 2 to 7 days following mixing of the test substance into the soil and incubation, with the same total transformation time as (2). A list of standard tests is provided in Fairbrother et al. (2002) and Spurgeon et al. (2005), together with recommendations on test soils and addition of test substances.

An alternative approach, based on a weight of evidence method, is to use appropriate published soil-quality guideline values. Such values, which should ideally be specifically aimed at determining threshold toxicity levels and should not be values derived for cleanup or sludge disposal, have been developed in the form of criteria or guidelines by a variety of jurisdictions (Table 6.7). These values may include assessment of background values or of added metals. For instance, The Netherlands Target and Intervention Values (Swartjes 1999; Rutgers and Den Besten 2005) represent threshold concentrations at which 95% and 50% of species are protected, respectively (the HC_5 and HC_{50}), and were developed using a species sensitivity distribution approach (SSD) that partly integrates the varying bioavail-

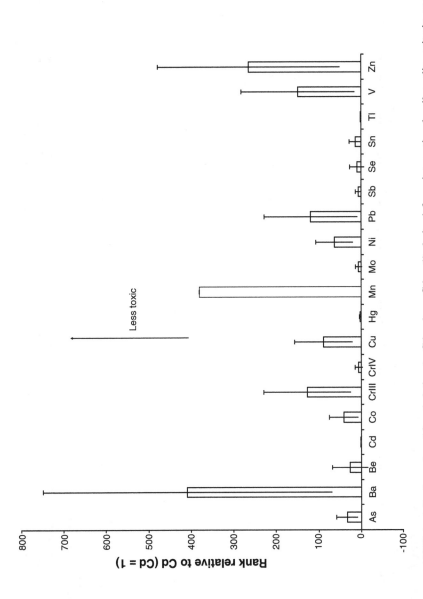

FIGURE 6.5 Average (±SD) toxicity hazard ranking (relative to Cd, where Cd = 1) derived from international soil quality criteria and guidelines (Table 6.7).

TABLE 6.7
Threshold Toxicity Values for Metals in Soils

	U.S. ECO-SSL	Denmark	Sweden	Finland	The Netherlands HC$_5$	The Netherlands HC$_{50}$	Belgium (Flanders)	Germany[a]	Switzerland	Czech Rep.	E. Europe[b]	Ireland	Canada
Ag													
As	31		15		34	85							12
B													
Ba	330				165	890							750
Be	30				1.1								4
Cd	0.4	0.3	0.4	0.3	0.8	13	2	0.4	0.8	0.4	2	1	1.4
Co	32		30		9	180							40
CrIII	7.9	50	120	80	100	220	130	30	75	130	90	100	64
CrIV			5										0.4
Cu	54	30	100	32	36	96	200	20	50	70	55	50	63
Fe													
Hg		0.1	1	0.2	0.3	36	10	0.1	0.8	0.4	2.1	1	6.6
Mn	152												
Mo					0.5	190							5
Ni	48	10	35	40	35	100	100	15	50	60	85	30	50
Pb	15	40	80	38	85	580	200	40	50	70	32	50	70
Sb	0.3				3.5								20
Se	1				0.7								1
Sn					19								5
Tl					1								1
V			120		42								130
Zn	120	100	350	90	140	350	600	60	200	150	100	150	200

[a] German values for sandy soil.
[b] E. Europe = Russia, Ukraine, Moldavia, and Belarus.

Source: The Netherlands: Swartjes FA. 1999. Risk Anal 19:1235-1249. USA EcoSSLs: www.epa.gov/ecotox/ecossl/. Denmark: Scott-Fordsman JJ. et al. 1996. Toxicol Ecotoxicol News 3:20–24. Canada: www.epa.gov/ecotox/ecossl/. Ireland: www.epa.ie/TechnicalGuidanceandAdvice/GuidanceDocuments/SoilQuality/. Sweden: www.internat. naturvardsverket.se/index.php3?main=/documents/legal/assess/assedoc/forstdoc/heavymet.htm. Belgium: www.ovam.be.

ability of added metals across soils into species variability (Posthuma et al. 2002). In contrast, the U.S. Environmental Protection Agency (EPA) (2003) approach to derive Ecological Soil Screening Levels (ECO-SSLs) uses data derived from conditions leading to high bioavailability, so that a geometric mean no-observed-effect concentration (NOEC) is used to derive the ECO-SSLs. In some jurisdictions, threshold values may include political considerations — for example, the inability of soils in that region to meet ecologically-derived thresholds due to past contamination — so that values are set at levels that are more achievable. Hazard rankings determined by relative international threshold regulatory values are most robust where these threshold values have been developed as part of a comprehensive risk assessment process (EPA 1998 — for example, the EPA ECO-SSLs and The Netherlands HC_{50} values).

Metal threshold values from Table 6.7 are ranked relative to Cd (=1) in Figure 6.5, so that metals having higher thresholds than Cd (that is, less toxic in soil) have values greater than 1, and metals where assessments indicated toxicity greater than Cd have values less than 1. This ranking, which only uses generic toxicity threshold values, does not account for the relative amounts of each contaminant entering the environment from anthropogenic sources and assumes that the threshold values have an ecological basis — which may not be true. For example, it is questionable whether Hg in a soils context is as toxic as appears from the low values and ranking in Table 6.7 and Figure 6.5.

Validation of this threshold-ranking approach can only be achieved through comparison of the relative toxicity of several metals to soil organisms among several soils and endpoints. Several studies have undertaken such a comparison but usually only for a few metals and a limited number of soils. For example, Doelman and Haanstra (1989) compared the effects of Cd, Cu, Ni, Pb, and Zn on soil enzyme activities in several soils. Their ranking of metals changed, depending on the soil type and according to the end point chosen, with the overall ranking being Cd > Cu > Zn > Ni > Cr > Pb. The variability in ranking casts doubt on the validity of a threshold-ranking approach for hazard assessment.

6.6 CONCLUSIONS AND RECOMMENDATIONS

Ranking of critical loads of metals differ, depending on the consideration of either steady-state concentrations in soil or a fixed time frame for the emissions. Time to reach steady state can exceed 10,000 years for some metals with high K_Ds. The relevance of predictions at such time frames is questionable, given that soil conditions often change over periods of years or decades. Therefore, it is recommended that critical loads be calculated on a fixed time frame, the choice of which is based on policy rather than science.

BCF/BAFs are not useful for ranking the hazard posed by metals, due to homeostasis for essential metals or nonlinear concentration-uptake relationships for nonessential metals, and because food concentration does not provide any indication of relative toxicity to either predators or prey organisms. Ranking in terms of bioaccumulation should be based on the relative values of BAFs and dietary toxicity reference values for wildlife. In terms of toxicity, the relative sensitivity of plants, microbes, and invertebrates appears to vary among metals; a large variation in metal-toxicity ranking among soils is observed even in identical studies.

Hazard ranking of metals in soil depends on the soil type and the toxicological pathways considered, that is, direct toxicity or considerations of secondary poisoning. Soil-based critical concentrations have been derived for 15 metals in the United States. and have been proposed as a weight-of-evidence approach for an initial ranking. Hazard ranking is possible using existing soil quality criteria and guidelines from various countries, but significant variation in relative rankings is evident. Most of these values are based on direct toxicity pathways, so that ranking using an average value across jurisdictions does not give equal weight to secondary poisoning issues. Further, comparison of hazard ranking using soil quality criteria and guidelines often does not correlate with hazard ranking in a single soil with a single test, so that ranking depends on the critical pathways considered (mammalian, microbial, plant, and so forth). A scaling system is needed to weight the importance of various pathways as part of any ranking system.

A better ranking system would involve actual toxicity tests using three different trophic levels under set conditions as recommended in Section 6.5. The best possible ranking system would include such toxicity testing in a weight of evidence assessment, including possible secondary poisoning. Future studies should focus on development of a bioavailability theory that can be integrated into models such as the BLM to assist in hazard ranking of metals in soils and reduce the uncertainty produced by abiotic factors affecting metal exposure. Finally, reliable methods need to be developed to determine adverse effects of metals on soil microbial processes or functions.

REFERENCES

Allen H, editor. 2002. Bioavailability of metals in terrestrial ecosystems: importance of partitioning for bioavailability in invertebrates. Pensacola, FL: SETAC Press.

Berkelaar EJ, Hale BA. 2003a. Accumulation of cadmium by durum wheat roots: bases for citrate-mediated exceptions to the free ion model. Environ Toxicol Chem 22:1155–1161.

Berkelaar EJ, Hale BA. 2003b. Cd Accumulation by durum wheat roots in ligand-buffered hydroponic culture: uptake of Cd-ligand complexes or enhanced diffusion? Can J Bot 81:755–763.

Beyer WN, Heinz GH, Redmon-Norwood AW. 1996. Environmental contaminants in wildlife: interpreting tissue concentrations. Boca Raton, FL: Lewis Publishers, p. 494.

Chan DY, Black W, Hale B. 2000. Bioaccumulation of cadmium from durum wheat diets in the livers and kidneys of mice. Bull Environ Contam Toxicol 64:526–533.

Chan DY, Fry N, Waisberg M, Black WD, Hale B. 2004. Kidney and liver accumulation of Cd from lettuce in the diet of rabbits: comparison of soluble-salt amended and plant-incorporated diets. J Toxicol Environ Health 67:397–411.

Chapman PM, McDonald BG, Lawrence GS. 2002. Weight of evidence frameworks for sediment quality and other assessments. Human Ecol Risk Assess 8:1489–1515.

Chapman PM, Wang F, Janssen C, Goulet RR, Kamunde CN. 2003. Conducting ecological risk assessments of inorganic metals and metalloids — current status. Human Ecol Risk Assess 9:641–697.

Crommentuijn T, Polder MD, Van de Plassche EJ. 1997. Maximum permissible concentrations and negligible concentrations for metals, taking background concentrations into account. National Institute of Public Health and the Environment, The Netherlands. Report nr 601501 001. (http://www.rivm.nl/bibliotheek/rapporten/601501001.html), p. 46.

De Groot AC, Peijinenburg WJGM, van der Hoop MAGT, Ritsema R, van Veen RPM. 1998. Heavy metals in Dutch field soils: an experimental and theoretical study on equilibrium partitioning. National Institute of Public Health and the Environment, The Netherlands. Report nr 607220 001. (http://www.rivm.nl/bibliotheek/rapporten/607220001.html), p. 260.

De Ruiter PC, Neutel A-M, Moore JC. 1995. Energetics, patterns of interaction strengths and stability in real ecosystems. Science 269:1257–1260.

Degryse F, Buekers J, Smolders E. 2003. Radiolabile cadmium and zinc in soil as affected by pH and source of contamination. Eur J Soil Sci 55:113–121.

Doelman P, Haanstra L. 1989. Short-and long-term effects of heavy metals on phosphatase activity in soils: an ecological dose-response model approach. Biol Fertil Soils 8:235–241.

Fairbrother A. 2003. Lines of evidence in ecological risk assessment. Human Ecol Risk Assess 9:1475–1491.

Fairbrother A, Glazebrook PW, van Straalen NM, Tarazona JV (eds). 2002. Test methods for hazard determination of metals and sparingly soluble metal compounds in soils. Pensacola, FL: SETAC Press, p. 97.

Hamon RE, McLaughlin MJ, Naidu R, Correll R. 1998. Long-term changes in cadmium bioavailability in soil. Environ Sci Technol 32:3699–3703.

Heikens A, Peijnenburg WJGM, Hendriks AJ. 2001. Bioaccumulation of heavy metals in terrestrial invertebrates. Environ Pollut 113:385–393.

Hindmarsh JT, McCurdy RF. 1996. Environmental aspects of arsenic toxicity. Crit Rev Clin Lab Sci 33:315–347.

Hough RL, Tye AM, Crout NML, McGrath SP, Zhang H, Young SD. 2005. Evaluating a 'free ion activity model' applied to metal uptake by *Lolium perenne* L. grown in contaminated soils. Plant Soil 270:1–12.

Johnson D, MacDonald D, Hendershot W, Hale B. 2003. Metals in northern forest ecosystems: role of vegetation in sequestration and cycling, and implications for ecological risk assessment. Human Ecol Risk Assess 9:749–766.

Leonard TL, Taylor GE, Gustin MS, Fernedez GCJ. 1998. Mercury and plants in contaminated soils: 1. Uptake, partitioning, and emission to the atmosphere. Environ Toxicol Chem 17:2063–2071.

Ma YB, Lombi E, Nolan A, McLaughlin MJ. 2006. Short-term natural attenuation of copper in soils: effects of time, temperature and soil characteristics. Environ Toxicol Chem 25:652–658.

McLaughlin MJ, Hamon RE, Parker DR, Pierzynski GM, Smolders E, Thornton I, Welp G. 2002. Soil chemistry. In: Fairbrother A, Glazebrook PW, van Straalen NM, Tarazona JV, editors. Test methods to determine hazards of sparingly soluble metal compounds in soils. Pensacola, FL: SETAC Press, p. 5–16.

Moolenaar SW, Temminghoff EJM, De Haan FAM. 1998. Modeling dynamic copper balances for a contaminated sandy soil following land use change from agriculture to forestry. Environ Pollut 103:117–125.

Oorts K, Ghesquiere U, Swimmer K. 2006. Discrepancy of the microbial response to elevated Cu between freshly spiked and long-term contaminated soils. Environ Toxicol Chem 26:229–237.

Parker DR, Pedler JF. 1997. Reevaluating the free-ion activity model of trace metal availability to higher plants. Plant Soil 196:223–228.

Posthuma L, Schouten T, Van Beelen P, Rutgers M. 2001. Forecasting effects of toxicants at the community level: four case studies comparing observed community effects of zinc with forecasts from a generic ecotoxicological risk assessment method. In: Rainbow PS, Hopkin SP, Crane M, editors. Forecasting the environmental fate and effects of chemicals. West Sussex, UK: Wiley and Sons. p.151–176.

Posthuma L, Suter GW, Traas TP, editors. 2002. Species sensitivity distributions in ecotoxicology. Boca Raton, FL: CRC Press.

Rutgers M, Den Besten P. 2005. The Netherlands perspective — soils and sediments. In: Thompson KC, Wadhia K, Loibner AP, editors. Environmental toxicity testing. Oxford, UK: Blackwell Publishing CRC Press, p. 269–289.

Sample BE, Beauchamp JJ, Efroymson RA, Suter GW II. 1998a. Development and validation of bioaccumulation models for small mammals. ES/ER/TM-219. Oak Ridge, TN: U.S. Department of Energy, Oak Ridge National Laboratory.

Sample BE, Beauchamp JJ, Efroymson RA, Suter GW II, Ashwood TL. 1998b. Development and validation of bioaccumulation models for earthworms. ES/ER/TM-220. Oak Ridge, TN: U.S. Department of Energy, Oak Ridge National Laboratory.

Scott-Fordsman JJ, Pedersen MB, Jensen J. 1996. Setting a soil quality criterion. Toxicol Ecotoxicol News 3:20–24.

Smolders E, McLaughlin MJ. 1996. Chloride increases Cd uptake in Swiss chard in a resin-buffered nutrient solution. Soil Sci Soc Am J 60:1443–1447.

Smolders E, Buekers J, Oliver I, McLaughlin MJ. 2004. Soil properties affecting microbial toxicity of zinc in laboratory-spiked and field-contaminated soils. Environ Toxicol Chem 23:2633–2640.

Spurgeon DJ, Svendson C, Hankard PK. 2005. Biological methods for assessing potentially contaminated soils. In: Thompson KC, Wadhia K, Loibner AP, editors. Environmental toxicity testing. Oxford, UK: Blackwell Publishing CRC Press, p. 163–194.

Swartjes FA. 1999. Risk-based assessment of soil and groundwater quality in The Netherlands: standards and remediation urgency. Risk Anal 19:1235–1249.

Thakali S, Allen HE, Di Toro D, Ponizovsky AA, Rooney C, Zhao FJ, McGrath S. 2005. Developing a TBLM: copper effect on barley root elongation. Abstract. Proceedings of the 8th International Conference on the Biogeochemistry of Trace Elements, April 3–7, 2005, Adelaide, Australia: CSIRO.

Thayer JS. 2002. Biological methylation of less-studied elements. Appl Organometal Chem 16:677–691.

USDOE (U.S. Department of Energy). 1998. Empirical models for the uptake of inorganic chemicals from soil by plants. BJC/OR-133. Oak Ridge, TN: Oakridge National Laboratory.

USEPA (U.S. Environmental Protection Agency). 1993. Exposure factors handbook. Available from: http://cfpub.epa.gov/ncea/cfm/wefh.cfm?ActType=default.

USEPA. 1998. Guidelines for ecological risk assessment. Washington, D.C.: Risk Assessment Forum. 175 p.

USEPA. 1999. Screening level ecological risk assessment protocol for hazardous waste combustion facilities. Available from: www.epa.gov/earth1r6/6pd.rcra_c/protocol/slerap2.htm.

USEPA. 2003. Guidance for developing ecological screening levels. Office of Solid Waste and Emergency Response. OSWER Directive 9285.7-55. Available from: http://www.epa.gov/ecotox/ecossl/SOPs.htm.

Weng L, Lexmond T, Wolthoorn A, Temminghoff E, Van Riemsdijk W. 2003. Phytotoxicity and bioavailability of nickel: chemical speciation and bioaccumulation. Environ Toxicol Chem 22:2180–2187.

Weng L, Lexmond T, Wolthoorn A, Temminghoff E, Van Riemsdijk W. 2004. Understanding the effects of soil characteristics on phytotoxicity and bioavailability of Ni using speciation models. Environ Sci Technol 38:156–162.

Young S, Crout N, Hutchinson J, Tye A, Tandy S, Nakhone L. 2006. Isotopic dilution methods. In: Hamon RE, McLaughlin MJ, editors. Natural attenuation of trace element availability in soils. Pensacola, FL: SETAC Press.

Appendix A: A Unit World Model for Hazard Assessment of Organics and Metals

A.1 THE AQUIVALENCE APPROACH

A brief outline is given here of the aquivalence approach; full details can be obtained in Mackay (2001) and Diamond et al. (1992). Concentrations are replaced and expressed by the equilibrium criterion of fugacity, which is a partial pressure or escaping tendency with units of pressure (Pa). Concentration C (mol/m^3) is linearly related to fugacity f (Pa) by a proportionality constant Z (mol/m^3·Pa) such that C is equal to Zf. When a substance achieves equilibrium between two phases such as soil (s) and water (w), a common fugacity applies, thus C_s/C_w, which is a dimensionless partition coefficient and is clearly also Z_s/Z_w. Definition of Z starts in the atmosphere where Z_A is 1/RT and proceeds to other phases using empirical or estimated partition coefficients, for example, Z for water (Z_W) is Z_A/K_{AW}, where K_{AW} is the air–water partition coefficient.

For nonvolatile substances such as most metals and ions, K_{AW} is 0; thus, Z_W becomes infinite. In reality, the partial pressure or fugacity is 0. In such cases, it is convenient to start the definition of Z_W in the aqueous phase where it is arbitrarily assigned a value of 1.0 and is dimensionless. If K_{AW} is 0, Z_A thus becomes 0, and Z_S is $Z_W K_{SW}$ or simply K_{SW} as before. The equilibrium criterion is then the *aquivalence* and can be designated A such that C = AZ. A has, thus, the dimensions of mol/m^3. Effectively, the expression Zf is both multiplied and divided by H, the Henry's law constant, to become (ZH)·(f/H), where ZH is the new dimensionless capacity term and (f/H) is the aquivalence or aquivalent concentration. As Z is 1.0 in aqueous solution, C and A are equal and are the actual dissolved concentration.

Processes of transport and transformation are expressed using D values, which are rate coefficients such that the rate N mol/h is DA. D thus has units of m^3/h and can be viewed as an equivalent flow rate. D can express advection, diffusion, and reaction processes in terms of flow rates, mass transfer coefficients, and areas or diffusivities; areas and path lengths; or reaction rate constants and volumes. Table A.1 summarizes these relationships for both fugacity and aquivalence formats.

TABLE A.1
D Values for Advective, Diffusive, and Reactive Processes

Type of Process	D Value	Definitions
Advection	$D = g \cdot Z$	g = flow rate (m³/h)
		Z = fugacity capacity
Diffusion	$D = A \cdot MTC \cdot Z$	A = area of boundary layer (m²)
		MTC = mass transfer coefficient of the boundary layer
Reaction	$D = k \cdot V \cdot Z$	k = reaction rate constant (h¹)
		V = volume of compartment

A.2 UNIT WORLD PARAMETERS

The Unit World here has dimensions and properties as listed in Table A.2. It is an area of 100,000 km² and represents an area similar to that of Ohio or Greece. The areas, depths, volumes, and compositions of the compartments can be varied, but for hazard ranking, a single set of values should be used. The equations used to define compartment Z and D values are listed in Table A.3 and Table A.4, respectively.

TABLE A.2
Dimensions and Properties of the Unit World

Compartment	Sub-compartment	Volume (m³)	Volume Fraction θ	Depth (m)	Area (m²)	Density (kg · m⁻³)	Organic Carbon Fraction
Air	Bulk	10^{14}	1	1000	10^{11}	1.19	—
	Air phase	10^{14}	1	—	—	1.19	—
	Aerosol	2000	2×10^{11}	—	—	2000	—
Water	Bulk	2×10^{11}	1	20	10^{10}	1000	—
	Water phase	2×10^{11}	1	—	—	1000	—
	Suspended solid sediment	1×10^{7}	5.0×10^{6}	—	—	1500	0.2
Soil	Bulk	1.8×10^{10}	1	0.2	9×10^{10}	1500	—
	Air phase	3.6×10^{9}	0.2	—	—	1.19	—
	Water phase	5.4×10^{9}	0.3	—	—	1000	—
	Solid phase	9.0×10^{9}	0.5	—	—	2400	0.02
Deep soil	Bulk	1.8×10^{11}	1	2.0	9×10^{10}	1500	—
	Air phase	3.6×10^{10}	0.2	—	—	1.19	—
	Water phase	5.4×10^{10}	0.3	—	—	1000	—
	Solid phase	9.0×10^{10}	0.5	—	—	2400	0.02
Sediment	Bulk	5.0×10^{8}	1	0.05	10^{10}	1280	—
	Water phase	4.0×10^{8}	0.8	—	—	1000	—
	Solid phase	1.0×10^{8}	0.2	—	—	2400	0.04
Deep sediment	Bulk	5.0×10^{9}	1	0.5	10^{10}	1280	—
	Water phase	4.0×10^{9}	0.8	—	—	1000	—
	Solid phase	1.0×10^{9}	0.2	—	—	2400	0.04

TABLE A.3
Z Values for the Environmental Media

Phase	Z	Fugacity Formulation	Aquivalence Formulation
Air	1	$1/RT$	H/RT
Water	2	$Z_1/K_{AW} = 1/H = C^S/P^S$	1
Soil solids	3	$Z_2\rho_3\phi_3 K_{OC}/1000$ or $Z_2\rho_3 K_{D3}/1000$	$Z_2 K_{Soil-W}$ or K_{Soil-W}
Sediment solids	4	$Z_2\rho_4\phi_4 K_{OC}/1000$ or $Z_2\rho_4 K_{D4}/1000$	$Z_2 K_{SedSolid-W}$ or $K_{SedSolid-W}$
Suspended sediment	5	$Z_2\rho_5\phi_5 K_{OC}/1000$ or $Z_2\rho_5 K_{D5}/1000$	$Z_2 K_{SusSolid-W}$ or $K_{SusSolid-W}$
Bulk sediment	6	$Z_4\theta_4 + Z_5\theta_5$	
Aerosol	7	$Z_1 K_{QA}$	
Deep soil solids	8	Z_3	$Z_2 K_{SoilDeep-W}$ or $K_{SoilDeep-W}$
Deep sediment (bulk)	9	Z_6	$Z_2\theta_2 + K_{SedSolidDeep-W}\theta_4$

TABLE A.4
D Values for the Transfer Processes

Environmental Media From	To	Process	Equation
Air	Water	Total	$D_{12} = D_{VW} + D_{RW} + D_{QW}$
		Absorption	$D_{VW} = A_W /(1/U_1 Z_1 + 1/U_2 Z_2)$
		Rain dissolution	$D_{RW} = U_3 A_W Z_2$
		Aerosol deposition	$D_{QW} = U_4 A_W Z_6$
Water	Air	Volatilization	$D_{21} = D_{VW}$
Air	Soil	Total	$D_{13} = D_{VS} + D_{RS} + D_{QS}$
		Absorption	$D_{VS} = 1/[1/D_S + 1/(D_W + D_A)]$
		Air boundary layer	$D_S = U_7 A_S Z_1$
		Air phase diffusion in soil	$D_A = U_5 A_S Z_1$
		Water phase diffusion in soil	$D_W = U_6 A_S Z_2$
		Rain dissolution	$D_{RS} = U_3 A_S Z_2$
		Deposition	$D_{QS} = U_4 A_S Z_6$
Soil	Air	Volatilization	$D_{31} = D_{VS}$
Water	Sediment	Diffusion plus deposition	$D_{25} = D_{WS} + D_D = U_8 A_W Z_2 + U_9 A_W Z_5$
Sediment	Water	Diffusion plus resuspension	$D_{52} = D_{WS} + D_R = U_8 A_W Z_2 + U_{10} A_W Z_4$
Soil	Water	Runoff of water and solids	$D_{32} = D_{SW} + D_{SS} = U_{11} A_S Z_2 + U_{12} A_S Z_3$
Soil	Deep soil	Transfer to lower layer	$D_{34} = U_{13} A_S Z_3$
Deep soil	Soil	Transfer to upper layer	$D_{43} = U_{13} A_S Z_4$
Sediment	Deep sediment	Transfer to lower layer	$D_{56} = U_{14} A_W Z_5$
Deep sediment	Sediment	Transfer to upper layer	$D_{65} = U_{14} A_W Z_6$

A.3 MASS BALANCE EQUATIONS

The six simultaneous differential equations describing the dynamic conditions are listed as follows. They are all linear in aquivalence, with M_i equal to the total chemical amount (mol) in the compartment. E describes emissions.

$$\frac{dM_1}{dt} = E_1 + A_2 D_{21} + A_3 D_{31} - A_1 D_{T1} \tag{A.1}$$

$$\frac{dM_2}{dt} = E_2 + A_1 D_{12} + A_3 D_{32} + A_4 D_{42} + A_5 D_{52} - A_2 D_{T2} \tag{A.2}$$

$$\frac{dM_3}{dt} = E_3 + A_1 D_{13} + A_4 D_{43} - A_3 D_{T3} \tag{A.3}$$

$$\frac{dM_4}{dt} = E_4 + A_3 D_{34} - A_4 D_{T4} \tag{A.4}$$

$$\frac{dM_5}{dt} = E_5 + A_2 D_{25} + A_6 D_{65} - A_5 D_{T5} \tag{A.5}$$

$$\frac{dM_6}{dt} = E_6 + A_5 D_{56} - A_6 D_{T6} \tag{A.6}$$

where:

$$D_{T1} = D_{13} + D_{A1} + D_{R1} + D_{12}$$

$$D_{T2} = D_{21} + D_{25} + D_{R2} + D_{A2}$$

$$D_{T3} = D_{31} + D_{32} + D_{34} + D_{R3}$$

$$D_{T4} = D_{43} + D_{42} + D_{R4} + D_{A4}$$

$$D_{T5} = D_{56} + D_{52} + D_{R5}$$

$$D_{T6} = D_{65} + D_{46} + D_{R6}$$

The 6 steady-state equations are obtained by setting the derivatives in Equation A.1 to Equation A.6 to 0 and rearranging as follows.

$$E_1 + A_2 D_{21} + A_3 D_{31} = A_1 D_{T1} \tag{A.7}$$

$$E_2 + A_1 D_{12} + A_3 D_{32} + A_4 D_{42} + A_5 D_{52} = A_2 D_{T2} \tag{A.8}$$

$$E_3 + A_1 D_{13} + A_4 D_{43} = A_3 D_{T3} \tag{A.9}$$

$$E_4 + A_3 D_{34} = A_4 D_{T4} \tag{A.10}$$

$$E_5 + A_2 D_{25} + A_6 D_{65} = A_5 D_{T5} \tag{A.11}$$

$$E_6 + A_5 D_{56} = A_6 D_{T6} \tag{A.12}$$

In the interests of brevity, the D values for all loss processes from a compartment are combined in a single total loss D value, designated D_{Ti}.

These 6 algebraic equations can be solved (with some difficulty) to obtain the aquivalence in water, then the other aquivalences as shown in Figure A.1. Inspection of this figure shows that the aquivalence is the sum of contributions for each of the 6 possible emissions. In reality, not all emissions are likely to apply, but they are included for completeness.

An advantage of this approach is that the equations characterizing the multitude of processes remain tractable and can be inspected for reasonableness. If expressed in concentration terms, the equations become very lengthy and the probability of introducing errors of transcription increases.

$$A_2 = \frac{E_2 + \dfrac{E_1 D_X}{J_3} + \dfrac{E_3}{J_3}\left(\dfrac{D_{31}J_{12}}{J_1} + D_{32} + \dfrac{D_{34}D_{42}}{D_{T4}}\right) + \dfrac{E_4 D_{43}}{D_{T4}J_3}\left(\dfrac{D_{31}J_{12}}{J_1} + D_{32} + \dfrac{D_{34}D_{42}}{D_{T4}}\right) + \dfrac{D_{42}E_4}{D_{T4}} + \dfrac{E_5 D_{52}}{J_5} + \dfrac{E_6 D_{65}D_{52}}{D_{T6}J_5}}{D_{T2} - \dfrac{D_{21}D_{12}}{J_1} - \dfrac{D_{21}D_{13}D_{32}}{J_1 J_3} - \dfrac{D_{25}D_{52}}{J_5} - \dfrac{D_{21}D_{13}D_{42}D_{34}}{J_1 J_3 D_{T4}}}$$

$$A_1 = \frac{E_1}{J_1} + \frac{E_3 D_{31}}{J_1 J_3} + \frac{E_4 D_{43}D_{31}}{D_{T4}J_3 J_1} + \frac{A_2 D_{21}}{J_1}$$

$$A_3 = \frac{E_3}{J_3} + \frac{E_4 D_{32}}{D_{T4}J_3} + \frac{A_1 D_{13}}{J_3}$$

$$A_4 = \frac{E_4 + A_3 D_{34}}{D_{T4}}$$

$$A_5 = \frac{E_5 + \dfrac{E_6 D_{65}}{D_{T6}} + A_2 D_{25}}{J_5}$$

$$A_6 = \frac{E_6 + A_5 D_{56}}{D_{T6}}$$

Substitutions

$$J_3 = D_{T3} - \frac{D_{34}D_{43}}{D_{T4}}$$

$$J_5 = D_{T5} - \frac{D_{56}D_{65}}{D_{T6}}$$

$$J_1 = D_{T1} - \frac{D_{13}D_{31}}{J_{34}}$$

$$J_2 = D_{T2} - \frac{D_{25}D_{52}}{J_{56}}$$

$$J_{12} = D_{12} + \frac{D_{13}D_{32}}{J_{34}} + \frac{D_{13}D_{42}D_{34}}{D_{T4}J_{34}}$$

FIGURE A.1 Steady-state solutions for the mass balance equations. Note that the aquivalence in water is solved first, then this value is substituted into the other equations. If, as is likely, not all emissions terms E_i apply, the corresponding terms that are products containing E_i can be ignored. The J terms are groups of D values with the dimensions of D values.

REFERENCES

Diamond ML, Mackay D, Welbourne PM. 1992. Models of multimedia partitioning of multispecies chemicals: the fugacity/aquivalence approach. Chemosphere 25:1907–1921.

Mackay D. 2001. Multimedia environmental models: the fugacity approach, 2nd ed. Boca Raton, FL: Lewis.

Index

A

Accumulation factor (ACF), 70, 71
ACF. *See* Accumulation factor
Acid volatile sulfide-binding, 23
Advection, D values for, 135, 136
Advective process, 33, 35–36, 38–40, 42–43, 135, 136
Air–water partition coefficient, 135
Algal tests, 95
Aluminum, 47
Antimony, 122, 127, 128
Aquatic systems, 2, 9, 48, 113. *See also* Water
Aquivalence, 135
Arsenic, 123, 124, 125, 127, 128

B

Bacterial degradation, 12
BAFs. *See* Bioaccumulation factors
Barium, 47, 127, 128
BCFs. *See* Bioconcentration factors (BCFs)
Beryllium, 127, 128
Bioaccumulation
    biodynamic models, 65–66, 75–78
    dietary toxicity and, 72
    hazard assessment and, 56–57
    models, 56–57, 75–78, 79–81
    potential, 124
        in soil invertebrates, 124
        in soil plants, 124
    regulatory objectives in hazard assessment, 56–57
    scientific basis of, 57–63
    in terrestrial food chain, 119–123
    toxicity and, 60, 62, 64
    UWM, 8
Bioaccumulation factors (BAFs)
    interpreting, 121
    inverse relationship with BCFs, 63–64
    measurement, 120–121
    in terrestrial systems, 119–120
    trophic transfer and, 65, 121–122
    uses, 65, 66
    values, 71–72

Bioavailability, 98–103
    aquatic models, 100
    in aquatic systems, 2
    construct, WHAM 5 in, 99
Bioconcentration factors (BCFs), 63–65
    chronic lethality and, 66–71
    interpreting, 121
    inverse relationship with BAFs, 63–64
    measurement, 120–121
    in terrestrial systems, 119–120
    trophic transfer and, 65, 121–122
    uses, 65, 66
    values, 71–72
Bioirrigation, 20
Biotic Ligand Model (BLM), 22, 95
    data gaps and future direction, 101
    development, 1, 99, 121
    framework mechanisms of, 99
    speciation component, 99
    uses, 95, 97, 101
Bioturbational transfer, 19
Bismuth, 122
BLM. *See* Biotic Ligand Model

C

Cadmium, 22, 45, 47, 66, 67, 68, 69
    ACF, 70
    bioaccumulation potential, 124
    chronic toxicity, 70
    dietary exposure, 98
    dissolved exposure, 79
    LBC, 66
    LC_{25}, 66
    methylation, 122
    sediment, 97
    soil enzymes and, 129
    toxicity, 66, 97
        chronic, 70
        dietary thresholds, 123
        hazard ranking, 127
        modification by uptake factors, 125
        threshold values, 128
    trophic transfer potential, 81
    uptake, 100
CBCs. *See* Critical body concentration

Milton Keynes UK
Ingram Content Group UK Ltd.
UKHW040053071024
449327UK00019B/521

9 780367 389550